THAT QUAIL, ROBERT

That Quail, Robert

by Margaret A. Stanger

with drawings by Cathy Baldwin

wm

WILLIAM MORROW

An Imprint of HarperCollins *Publishers*

A hardcover edition of this book was published in 1966 by J. B. Lippincott Co. It is here reprinted by arrangement with J. B. Lippincott Co.

HarperCollins books may be purchased for educational, business, or sales promotional use. For information, please call or write: Special Markets Department, HarperCollins Publishers, Inc., 10 East 53rd Street, New York, NY 10022. Telephone: (212) 207-7528; Fax: (212) 207-7222.

First HarperPerennial edition published 1992.

LIBRARY OF CONGRESS CATALOG CARD NUMBER 92-52614

ISBN 978-0-06-081246-1

30 29 28 27 26 25 24

To Mildred and Tommy Kienzle,
who provided the loving environment and understanding care which resulted in the development of a unique personality—Robert—and who generously permitted me to write this book

1

THE MIRACLE

UNTIL JULY 11, 1962, we had no hint of the change about to take place in our lives. On that date there was no Robert; there was just an abandoned egg in a deserted wet nest. We had known that there was a quail nesting in the deep grass beyond the rose garden. My friend and neighbor, Dr. Thomas Kienzle, had discovered the nest in June, and he had carefully left the surrounding grass unmowed.

By dint of patience and long periods of standing motionless, he had seen the little hen go to the nest and had watched her as she reached out with her bill and carefully covered herself with the grasses around her, from one side, from the other side, from front and from back, until she was completely hidden. She had chosen her spot well. Tommy and Mildred had built their house at Orleans, on Cape Cod, high on a bank above a lake, and since it was surrounded by extensive woods, it was a haven for man and bird. As he cleared lawns and paths, Tommy had purposely left piles of brush in among the trees, because quail had often been seen around the place. Of course he had no idea as to when the eggs would hatch, but both he and his wife were on the alert for them.

When their two boys were in their teens, Dr. and Mrs. Kienzle had spent several summers on Cape Cod, and the decision to make this their retirement home had been unanimous. Few men have had a better right to look forward to retirement. For thirty years he had been a member of the medical department of Standard Oil Company, New Jersey; during much of this time he had been responsible for the medical welfare of the thousands of men on their huge fleet of tankers. Previously he had been Assistant Surgeon in the United States Public Health Service. He can tell fascinating stories of his work in Mississippi, on flood control under President Hoover.

Since both the doctor and his wife grew up in Kentucky where they early acquired a deep-rooted love of wildlife, it is not strange that when the time came to locate on Cape Cod, they resisted the efforts of the real estate agent to interest them in seashore property, and chose instead four wooded acres on the high bank of what was originally Fresh Pond, but later known as Crystal Lake. The land was wild and rough, and much of it has been left that way, making it a haven for birds and animals of all kinds, including deer.

In a spacious clearing they built their attractive home, with floor-to-ceiling glass toward the wide patio overlooking the lake. The lawn extends from patio to steps going down the seventy feet to the shore and dock. Wild ducks, geese, herons and gulls are in abundance and the many bird feeders bring field glasses and bird books into frequent use. Into this lovely setting they moved, bringing with them Old Faithful, a goldfish whose actual age is unknown, but who has been with them sixteen years. It is a welcoming home, and I was delighted to have them here, and although I had never been a bird watcher, I soon found myself sharing their enthusiasm.

On that eventful July morning, Tommy went to get the car for a trip to Lexington. As he neared the garage, which is some distance from the house, he was aware of movement in the grass, and his attention was immediately drawn to a female quail apparently in distress. She was in the middle of the driveway, dragging one wing as though it were broken, and seeming to struggle. Tommy stood very still and then saw the male bird, going through the same performance. He realized immediately that he was witnessing one of the most remarkable acts of parental bravery known to the world. The parents were definitely drawing his attention to themselves, away from their young. As he stood quietly he saw two or three tiny balls of fluff moving off the driveway to safety, evidently obeying some kind of direction from the mother. Tommy moved softly back toward the house, and was rewarded by seeing twelve young quail led into the woods. In a few minutes he got the car, picked up Mildred and recounted what he had just been privileged to see. As they drove out of the yard he said:

"Look! There they are! The mother quail and—two, four, eight—there are twelve of them, just going past the rose bush toward the pine trees."

He and his wife sat quietly in the car as the mother quail led the little balls of brown fluff off to safety. Their coloring was so protective that it was hard to follow them even a short distance.

"I've read that the quail hen never returns to the place where the chicks were hatched, so when we come home we will have a look at the nest," said Mildred as they drove on down the driveway. But before they returned, there was a thunderstorm, and they did not visit the nest until the next morning. Even though they knew almost exactly where it was, it was cleverly hidden and not easy to find. They stood looking at it—just a little cup-shaped depression in among the grass.

"Wait a minute," said Tommy. "I think there is something in it." Sure enough, down in the mud were two eggs, one badly cracked and one perfect. He picked up the perfect one. It was very dirty and covered with tiny, lively mites. They took it into the house, washed it with cold water, sprayed it with bug spray and detergent and left it on the kitchen counter as a curiosity. Later in the day they noticed a small crack in the shell, so just in case, they put a small boudoir lamp beside it for warmth.

For two days they watched it. Nothing happened. But on the third day they thought the egg moved slightly. As they held it to their ears, they could hear a faint ticking inside, like a miniature time bomb.

A quail egg is a lovely thing. It is snowy white, about an inch long, softly rounded at one end and fairly pointed at the other. There certainly was something going on in this one. As they stood transfixed, tiny holes began to appear around the pointed end. Ornithologists tell us that the chick always comes out from the larger end of the egg, but not this one. When there was an almost complete circle of holes, a slight convulsive shudder came from within, and the shell parted. There emerged slowly something resembling a wet bumblebee in size and general appearance. It lay there apparently exhausted, and the doctor and his wife stood watching it in amazement. The living room clock interrupted the silence by striking two, and Mildred realized in horror that they had completely forgotten lunch. She set about preparing it while the doctor continued to gaze. The tiny thing was drying off perceptibly, moment by moment, and before they went in to lunch, they moved the lamp and

the chick to the corner of the counter and, more to keep it warm and protected than for any other reason, they barricaded it with cereal boxes and a bag of groceries. When halfway through the meal, they heard a tiny little chirp, and there in the doorway stood the baby, the body now fluffed to the size of an English walnut, tottering on fragile legs, balanced precariously on big feet. They rushed to pick it up, realizing that it must have fallen off the edge of the counter to the floor. A small space between a cereal box and the bag showed clearly where the bird had found its way out of the barricade. Even at the age of about an hour, the chick had followed the sound of human voices and found the first living creatures it was to encounter—two human beings.

Greatly concerned lest the fall had injured the infant, Tommy and Mildred feared this might be the end of the story—but it wasn't. It was just the beginning. It was evident that more security must be provided, and a carton was found and bird, boudoir lamp and all were placed inside it. It looked pitiful and miserably alone. Suddenly Mildred remembered a small lamb's-wool duster about a foot long, including the handle, which she put in the corner of the box, tying it so that the wooly part was an inch or so above the bottom of the box. That looked better. Bobby White, as he was immediately called, seemed to like it too, and was under it almost as soon as it was in place. Within half an hour from the time he was hatched, the baby had begun to change in appearance. He was now brown instead of black, and grew fluffier and downier by the minute.

What to do now was the question. They called a friend who was interested in birds, and received no encouragement at all as to their being able to save its life. However, she suggested that they go to the local duck farm and get some baby chick starter, which they immediately did. A Mason jar top full of chick starter, another containing water, and they were in business. The baby, while seeming to like the duster from the beginning, paid no attention to the food and water. Not for eight hours did he eat at all. Many people, hearing the tale, have assumed that he was force-fed at first. Not so. From the very beginning he was very self-sufficient. The only help he had was when his tiny bill was gently dipped in the water. From then on—no help.

As I attempt to chronicle this extraordinary tale, I am ever grateful that as a close friend of Mildred and Tommy I have been privileged to watch and share the life and love of this tiny bird as he grew and developed. I did not see him until the second morning of his life, when I was taken into the kitchen and invited to look in the carton. I saw the lamp, the duster and the receptacles of food and water, but nothing else. Mildred gently picked up the duster and there snuggled inside it was the exquisite little bird. She picked him out from the nest of wool and he tackled his breakfast with enthusiasm. I felt strongly that he deserved a more dignified name than Bobby, and I immediately called him Robert; and Robert he has been to all of us since then. Even then, at the puffball stage, he was really beautiful, with muted shadings in his coloring, lighter on the breast than on the back, and with very bright little eyes.

The lamb's-wool duster proved to have been a real inspiration. Later we discovered that had he stayed with his quail mother he would have snuggled in very similar comfort, as the underfeathers of the adult quail are so fine and soft that they feel much more like fur or wool than feathers.

I wonder why it is that people are so outspokenly gloomy about the chances for survival of a little wild infant. At this stage the prophecy of the neighborhood was, "He'll never live."

Then, after he had survived and thrived for several days, other calamity-howlers began bringing up another bugaboo: "You'll never be allowed to keep him" . . . "It's against the law to keep a wild bird" . . . "You'll be in trouble," and so on and on.

The doctor and his wife had very firm answers:

"We aren't going to cage this little thing. We aren't even handling him any more than we have to, to clean the carton and keep him warm and fed. He will be freed outdoors as soon as he is strong enough, but we can't put him out in the big world now. He wouldn't stand a ghost of a chance if we did."

It turned out that only Robert himself knew the answers to the crepe-hangers. In the first place, he had no intention of leaving his new home for the unknown, and time was to prove that he also knew the answer to the game warden question; for we were to learn that our idea that he might be taken back into his own family of brothers and sisters was a complete fallacy: quail will not accept any bird who has been in contact with human beings.

We should have realized earlier than we did that, far from having a bird in captivity, we were helplessly and hopelessly ensnared and enamored. Robert's development was spectacular in three areas: vocabulary, plumage and general personality traits. From the second day, he greeted his providers with distinctive chirps of pleasure and anticipation. The clucks while he was eating and scratching were in a rather high register and sweet in tone. Finely ground wild-bird food and fine gravel had been introduced, unmistakably to his delight. We could tell without looking in the carton what was going on—the happy, busy sounds and then little mournful coos which meant he wanted companionship and the indescribable little soft trills, more throaty purr than birdcall, when he was drowsy, dimin-

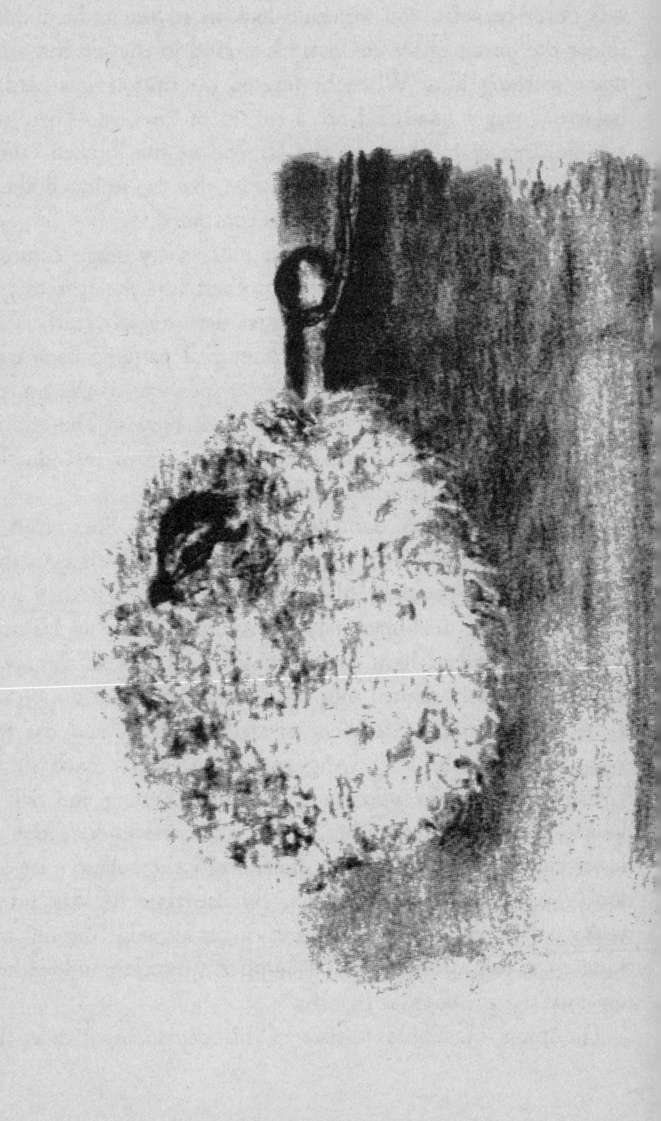

ishing in frequency and volume until at last he was asleep. He was never nervous, and we could look in at him as he cuddled under the duster or we could work around in the kitchen without disturbing him. When he uttered his trills it was hard to believe it was a quail and not a canary or lovebird. There was a large cage of lovebirds at the far end of the kitchen, and I often wondered if it could be possible that he imitated them. We shall never know, but the trills continued.

Robert omitted the unattractive pinfeathery stage common to the chicks of domestic fowl. His chest was the first to give indication of its subsequent beauty as the tiny soft feathers appeared over the down in a tiny chain-mail pattern. Each individual little beige feather had a lighter spot toward the tip, the tip itself being a fine border of very dark brown. The end result was exquisite, and almost defied the art of reproduction when an artist did Robert's portrait in oils.

The top of his head darkened soon, with tiny lines of an almost old-gold color down the sides which set off his eyes startlingly. We had supposed the eyes to be black, but they were dark brown with coal-black pupils. Around the base of his neck appeared a band of light brown which darkened at the lower edge, and in front melted into the general pattern of the breast. Tiny softly patterned feathers covered his back, and the last to appear were the long wing and tail feathers. Each single feather, with the exception of those forming wing and tail, is double, with the outer feather having a separate underfeather of down coming from the same shaft. Even the tiniest ones are double, giving perfect insulation. By the time he was a few weeks old he was a beautiful little quail, although he did not achieve the full measure of his magnificent coloring and feathering until the age of three months.

The most incredible feature of his development was the

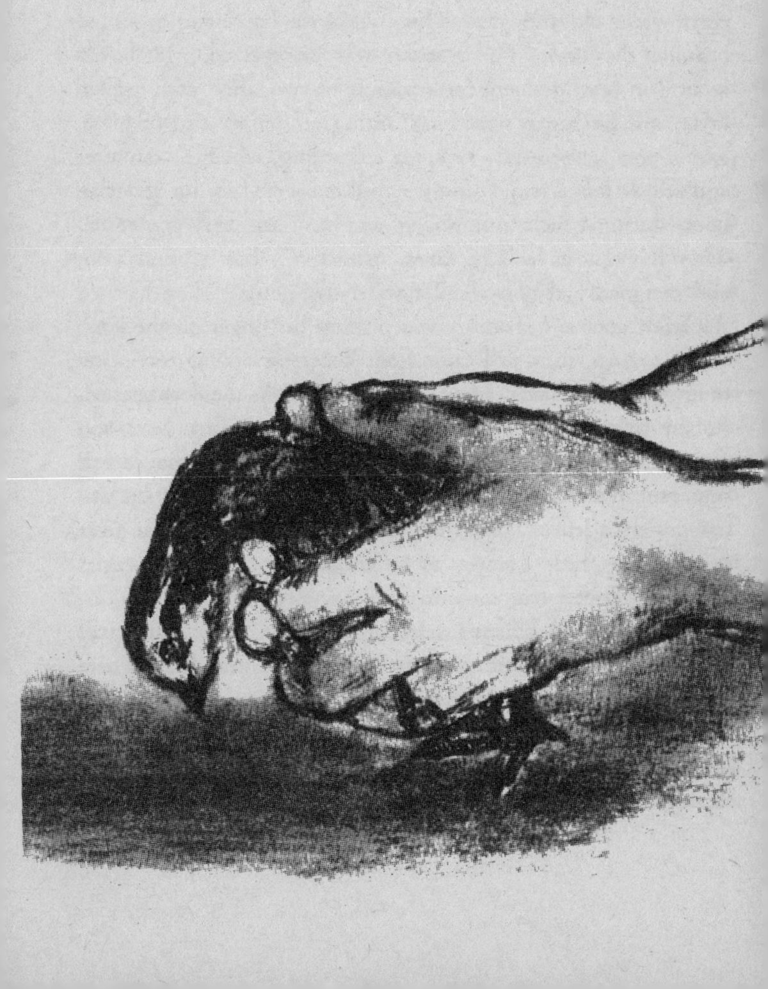

emerging personality. At the age of two weeks, he showed a real fondness for the doctor, often going to sleep while in his hands during the intervals while his carton was being cleaned. This was a clear indication that the time had come to let him free. The summing up of the discussion was "We have to do it some time. We just have to make up our minds to that."

On a warm, sunny day in early August when Robert was three weeks old, they carried him out of the house and deposited him on the lawn. The reaction was unexpected. He looked around in bewilderment for a minute or two, then with excited little calls he began scratching, biting off tender tips of grass, and almost immediately he spied a tiny bug, which he ran after and ate. Mildred and Tommy stayed out with him for about an hour, during which time Robert was busy and very contented. They felt that at least he knew instinctively how to find food, and resignedly they walked toward the house. They reached the front door and Tommy was putting his finger on the latch when a sharp, shrill call came from Robert several yards behind them, a call that said, "Hey! Wait for me!" As the door opened, Robert came running as fast as his legs could carry him, and darted in the house ahead of them. After this performance had been repeated for many days, the answer was evident: Robert was not going to leave. From then on, he was out of doors a great deal, staying near Tommy as he weeded flower beds or raked the lawn. Robert had come to stay.

On several of Robert's outdoor excursions, his own quail family, all twelve of them, were often quite near him. There was never the slightest sign of recognition, much less reunion, on either side. Who snubbed whom was not clear. Since the grounds around the house were liberally equipped with bird-feeders of every description, and since several patches of ground were well sprinkled with wild-bird food, it was not strange that the mother quail had kept her family nearby.

There is abundant evidence to show that the mother quail's injured-bird performance witnessed by the doctor was by no means uncommon. A friend of Robert's told me that he drove home one evening from a neighboring town when it was dark enough for him to have the headlights on. In the road ahead of him he saw a "dead" quail, lying in the road, with feathers all straggly and wings limply outstretched. Being a compassionate man, he did not want the little body run over repeatedly by other cars; he stopped and went to remove the poor creature. When he was within a few feet of her she moved, got up and revealed four young quail who had been hidden beneath her. With little low cries she herded them to the side of the road, where she joined a number of other chicks who had been standing there as still as the grass itself—no doubt in response to a directive from the mother, who, with the four little stragglers, had not quite made it across the road before the oncoming car.

From my picture window I have watched a mother quail with her covey of young scattered around her, and have seen her find a succulent bug or seed, which she does not eat herself. Instead, by means of a series of sharp taps on the ground with her bill, she summons the chicks, who come running—and how they can run. This bit of knowledge was most useful in our relationship to Robert. If a little spider or bug was seen on the floor, quick taps brought him scurrying from any room in the house to pounce on the tidbit.

From the moment Robert first refused the invitation to return to the wild and ran back into the doorway, the house was his. His adjustment to its size was quite dramatic. With neck outstretched and head held slightly to one side, and with very measured tread, he stalked around investigating everything. An early discovery was that sometimes a tiny spider could be found in the baseboard. For weeks, on being taken out of his carton in the morning, Robert would methodically make the rounds investigating baseboards. The house was large and quite new, and for some time those little spiders had been pests. Within the first few weeks Robert took care of that and completely rid the home of the spiders.

One of the surprises at his first becoming a regular member of the family was his marked liking for companionship. Where the family members were, there, too, was Robert. If Mildred was sewing in the sewing room, Robert was there, investigating patterns, running off with bits of cloth and generally participating in the project. If Tommy was reading the paper, Robert was in his lap, begging for attention. When he found that Tommy's interest was really on that paper, he would give up to the point of nestling in the crook of Tommy's arm, or on his shoulder, especially if he was wearing a soft woolen shirt.

Robert was highly sociable. The more people around, the merrier. In the living room there is a large circular coffee table in front of a semicircular davenport. Tea was often served there, served around Robert. He greeted every newcomer with cries of real delight. The cries were always the same, and continued while he investigated the caller's shoes. He had a distinct aversion to open-toed shoes or sandals of any kind. He liked old shoes best, especially those belonging to Tommy, and his preference of Mildred's footwear was a pair of flat-heeled red shoes. Once the shoe inspection was over and people settled down to visit or have tea, Robert got on the middle of the table and devoted himself unconcernedly to preening. He kept an eye on any cake or cookie which appeared, and daintily helped himself. Tea was good when it had cooled down.

It was amusing to observe the caution with which people approached Robert when they saw him for the first time. They were so careful to move slowly, not to make any sudden motions, to keep their voices down in order not to frighten him in any way. Frighten Robert? It couldn't be done. Experiencing nothing but love and kindness in his life, he developed a feeling of complete security and confidence toward any and all humans. I said Robert could not be frightened. One exception to this statement was his fear of birds. A sea gull—or even a sparrow—flying over him outdoors would send him into a state of panic. At such times he seemed to freeze, then crouch down and flatten himself against the ground, then he would dart under a nearby bush or big-leafed plant. He had no mother to teach him the danger of hawks; no bird ever attacked him; he ignored completely the lovebirds in the house, even when they were allowed to fly free for a while—but outdoors any bird terrified him.

We celebrated his two-months-old anniversary on September 14, 1962. We had tea out on the Kienzles' patio and for Robert I had a cupcake topped with two small candles, the rim of the bread and butter plate on which it was served being covered with fresh chickweed, the stems anchored by the cake. They were not anchored long. Every bit of it disappeared in a few minutes, and then he ate all the cake he could hold. We smile now to think what an important date that seemed. Later we celebrated only actual birthdays.

2

REGISTRATION

IN RETROSPECT, Mildred and Tommy realized that had they just known enough to leave things to Robert, they need not have been so concerned about letting him go. Meanwhile, they made plans to go to a bird sanctuary farther down the Cape and tell the whole story to the Audubon people. Mr. Wallace Bailey, the very able director of the sanctuary, was to give a lecture soon. When the time came, Mildred and Tommy attended. They saw the sanctuary, saw birds which had been brought there wounded or sick, which were living there happily in cared-for freedom; they were greatly impressed by Mr. Bailey, his ideas, his practices and his personality. They arranged to have a few minutes with him after the lecture, and he listened with real interest to the whole remarkable story. At the end of it Mildred summoned her courage and asked, "Do you think we should bring him down here to you?"

Mr. Bailey considered the matter for a moment and then said, "Well, to tell you the truth, we have not had much success with quail. It seems to me that, since the bird has done so well under your care, it might be better just to let things continue as they are. However, he must be banded for our records, and then if he leaves you, in the mating season for example, we would have the identification record."

Mildred and Tommy agreed readily, and asked what the banding would entail.

"There are some new bands coming out, very light plastic ones," said Mr. Bailey. "I'm going to Boston soon to get some of them. In the meantime we'd better put one of the present bands on." He showed them the band, a tiny, very light metal object with a number stamped on it.

The next day, while Robert was held in the safety of Tommy's hands, the little band was slipped on. A tiny clamp held it securely and it was just the right size, as Mr. Bailey had known it would be. It didn't bother Robert at all; he completely ignored it. Now he was legally registered and perhaps protected for the future. In these days of numeralization—zip codes, nine-digit telephone numbers, social security numbers and so on—it is only fitting and proper that Robert, having chosen to live with humans, should have a number of his own. Be it known henceforth that Robert became:

633-87201 Quail
U. S. Wildlife Service,
Washington, D.C.

Robert's new status of ornithological citizenship with all rights and privileges appertaining thereto, did not impress him at all. He ignored it as he ignored the little band on his left leg. I do not want to give the impression that Robert was insensitive to change. Physical changes in his immediate surroundings upset him. Let the position of a davenport be changed, or let the corner of a rug be inadvertently turned back, and he made it known that something was wrong. In case of the rug, he went into quite an act. One day he encountered the folded corner as he was running across the room. He skidded to a full stop, stretched his neck almost two inches longer than normal and, holding his head to one side, began stalking. Back and forth he went, uttering little cries of consternation. Someone heard him, restored the rug to its normal flatness and Robert, with a little "Well, that's taken care of" chirp, resumed his usual shape and continued on the original errand.

Displacement of very small objects did not escape his notice. He was intimately acquainted with the objects on Mildred's dressing table, since he was usually on it when she was sitting there. But let a bottle of nail polish be left where it did not belong, as on the stand in the bathroom, and Robert's world was disturbed. He would stalk and call, stalk and call, until the situation was remedied.

He seemed to accept the fact that human beings did not have special places where they could always be found. However, there were times when they should *be* in certain places, as for instance at the breakfast table. One of the enjoyments of retirement for the doctor and his wife was to have breakfast at leisure. Robert was always at the table too, having orange juice, bits of toast, sips of coffee, and occasionally some scrambled egg. One morning the doctor got up early and went out of doors to inspect some plants he had set out the day before. When Mildred got up, she noticed that she had really overslept, and thinking that Tommy would be hungry by then, she prepared breakfast, called Tommy, and then went into the bedroom to put on a dress instead of a housecoat. Tommy was at the table, Robert was at the table, but something was wrong. The other person was

Not at the table. Robert looked at Tommy, looked toward Mildred's accustomed place, looked back at Tommy, would not take his orange juice, paced back and forth and finally, with a little call of interrogation, flew down from the table, into the bedroom and pecked at Mildred's feet. He followed her into the dining room, flew up on the table, and life was normal again.

Speaking of bedrooms—at about the same time as the banding, Robert gave up sleeping in the carton. At bedtime, which for Robert was usually about seven o'clock unless there happened to be company whom he did not want to miss, he signified that he was ready to go to sleep by getting up in a philodendron plant, which was on an end table. There he would make his sleepy sounds, pull up one leg, settle himself on the other, and with eyes closed, his head would droop. He was unmistakably tired. Someone was always there to pick him up and put him in the carton under his lamp. But one night he disappeared. He was found later on a high shelf in the dressing room off the big bedroom, sound asleep on a red velvet pillbox hat. He looked so cozy that they left him there, and from then on that was it. The hat was put in a cellophane bag, which did not detract from its comfort at all. Once settled down, he was there for the night. I often went in to peek at him. Since the crown of the hat was soft, just his head was visible. When the light went on, he would open one eye, give a little purr of recognition, and go back to sleep.

I once heard a woman with twelve grandchildren ask a young mother if her little boy, then at the creeping stage, had fallen out of bed yet. The young mother was horrified at the very idea of such a thing. "Well," said the grandmother calmly, "he will. And it won't hurt him, but he will."

Robert, like other young creatures, passed that milestone.

One evening while the family were reading in the living room, they heard a little frightened cry, a fluttering of wings and a little thud. They rushed in and, sure enough, he had fallen out of bed. He had evidently gotten too close to the edge of the shelf, lost his footing and his balance and down he went, striking his head against a corner of a sewing machine, which was also kept in the closet, before he could recover his equilibrium.

They picked him up to find that the feathers on the very top of his head were gone, and with them a bit of his scalp. It bled profusely, but healed nicely by having ointment rubbed in immediately, and several times afterward. I think this is when he developed such a liking for having the top and side of his head rubbed. He would lie motionless in someone's hand as long as the rubbing continued.

After a second fall, which fortunately did not hurt him at all, a tiny night light was placed in the dressing room. We cannot shield any young thing from all possible vicissitudes in the growing-up process, and we all know the dangers inherent in overprotectiveness. But there is a distinction between overprotectiveness and common-sense caution, and it is the latter which we endeavored to maintain in rearing Robert.

"Up with the birds" did not apply to Robert. Although he insisted that both Tommy and Mildred come to breakfast when he was up, he did not reciprocate. He was a late sleeper, often missing breakfast with the family and not putting in an appearance till ten or eleven o'clock. His getting-up routine was most amazing. You might expect him to be a bit groggy on first waking, but he was not; he was in a hurry. He would fly down from the shelf and at a dead run dash (and that is the only word for it) into the bedroom, around the big dresser, and into the bathroom, where he made two, always two, large birdlike deposits on a piece of Kleenex placed on the floor for that purpose. Nothing could distract him on that run. While we are on this subject, in case there are raised eyebrows at the thought of a quail free in the house, let the eyebrows resume their normal position. After the morning evacuation, there were occasional tiny dry droppings like those of a parakeet. His favorite resting places were well known, as for instance the wide upholstered arms of the davenport, and these places were provided with nice-looking terry cloth pot-holders. A little shake into a wastebasket now and then took care of that. There were very, very few accidents.

As the first autumn season set in, there were gradual changes in Robert's plumage. Not only did his coat become heavier and richer; the dark feathers darkened further, accentuating the contrast in pattern. Also, on his topknot appeared little spines, resembling tiny plastic pins with the tips showing beyond the lovely brown and black head feathers. Their purpose was soon apparent: to raise the crest in moments of excitement or pleasure. The topknot would be erect as the bird ran across the patio in pursuit of a bug. It would also rise when he heard a new sound which for the moment he could not identify. When the whistling teakettle started to whistle, in fact before it was audible to human ears, the topknot stood up and Robert sounded off, usually running to the kitchen to get that noise turned off.

This was evidently not the response of fear. But certain sounds, especially in the house, caused him to fluff out to twice his normal size, every feather seeming to change position. We have a photograph of him taken on the desk when the telephone was ringing. He knew it should be answered. He enjoyed the telephone and, hopping up onto the shoulder of the person talking, he would keep up a chatter himself. Given the chance, he chirruped and sang into the mouthpiece, responding to the voice coming from the other end.

One day the telephone repairman came. He had read about Robert in the local paper, but still was far from prepared to find him examining each tool, practically standing on tiptoe to look in the tool case, talking all the time. When the repairman called the operator to check on the condition of the telephone, Robert hopped up on his shoulder ready for a visit. The man was enchanted. Of course we could hear only his end of the conversation, but what went on at the other end of the line was obvious. It went like this:

"Checking on 0017—no! No, there isn't anything on this line! Do you know what those sounds are? It's a quail! No, I'm not kidding. Honest, it's a quail. Wait a minute!" He held the mouthpiece up to Robert, who responded as he always did. "How about that!" Then, "Sure it's in the house. You don't think I'm calling from out in the woods, do you? It's that quail, Robert."

The man stayed several minutes after the work was done, just enjoying Robert, shaking his head in disbelief. He held Robert, he got down on the floor to watch him drinking and eating bird seed from his tray. It was a never-ending source of interest to watch people's reactions to Robert. Of the hundreds who called on him, only perhaps three or four failed to respond to the marvel of this friendly, bouncy, affectionate and sociable little bird—a member of a species noted for timidity and shyness.

3

A CHANGE OF SEASON

QUAIL do not migrate, and are native as far north as Maine, so I was surprised by Robert's behavior with the coming of winter. When the first snow fell, Robert was having none of it. The first time he saw it—as usual, he had gone to the door and asked to be taken out of doors—he was bewildered. Out stretched the neck, to one side went the head, and he walked back and forth along the edge of the protected part of the patio, not knowing what to make of this new look. He finally ventured forth a step or two, backing up immediately onto the familiar cement floor. After several such attempts and some vociferous complaining, he turned tail and rushed to the door to be taken in. At first I was ready to call Robert a sissy, but after all he was an indoor bird, accustomed to his creature comforts.

In about an hour he seemed to have forgotten the experience, and again asked for an outing. But the snow was still there. It was a light snow, and his careful steps in it left little quail footprints such as those of us who are fortunate enough to live in the country often admire on a snowy morning. This time when he came in, he evidently decided it was for good, and not again that day did he suggest another venture.

Everyone has seen a man standing before a fireplace, contentedly warming the seat of his trousers. Well! Robert went out to the kitchen, where his lamp was always lighted, but instead of settling under it as he usually did, he flew up on top of the lampshade, which brought him on a level with the windowsill, where, gently moving back and forth without moving his feet, he warmed his derriere. This proved so delightful that the shade became a favorite spot on snowy days. It was a comical sight to see him sitting there, comfortably squatting in a soft little lump over the lamp, watching through the window other birds braving the snow as birds should, happily getting the food put there for them. It would be presumptuous to say that Robert pitied those birds, but one thing is certain: he did not envy them. He had that smug, comfortable look one sees on a cat lying in front of a fireplace with paws tucked under.

Love of warmth seems to be universal, and Robert loved the fireplace when there was a blazing fire as much as the next person. He was always visibly excited when a fire was being laid, and had we not known that Robert knew how to take care of himself, we would have worried about his lack of caution. I have read that all wild animals and birds have an instinctive fear of flame. This was not true of Robert. He got on the hearth, right up next to the fire-screen, and not even the loudest crackling and popping of burning apple wood disturbed him in the least degree.

I once read a scientific article which discussed a number of experiments involving mother substitutes for animals. The one I remember most vividly was the one conducted with infant monkeys. The babies had been taken from their mothers and were divided into two groups of perhaps three monkeys each. In one cage, the group was provided with a mother substitute consisting of a wad of chicken wire covered with terry cloth, while the group in the other cage had nothing at all. One convincing demonstration showed that when danger and fear were introduced, the babies with the terry-cloth mother rushed to it and clutched it and evidently derived a feeling of security and safety from it; but the other group ran frantically around the cage in terror and without comfort.

The lamp seemed to represent a mother substitute for Robert. It was a small boudoir type, perhaps twelve inches high from top to bottom, with about five inches between the table top and the lower edge of the shade. He had received his first warmth from that lamp, and he continued to seem to love it. It was always kept in a corner on the kitchen counter, with seed and water nearby. It was an appealing sight, on a wintry night, to look in the kitchen window and see the pool of light with Robert underneath, sitting flattened out or, much more frequently, lying on his side with feet and legs and neck stretched out in complete relaxation.

His settling-down process was always the same. He would fly up to the counter, go to the lamp, then with just his body visible, walk round and round rubbing the inside of the shade with head and shoulder. Sometimes after a particularly busy day, if Robert had had an unusual number of callers, or if he had helped with the housecleaning by following the dustcloth over what must have seemed to him acres of polished surfaces, the lamp was brought to the dining room and put at one end of the long refectory table. On these occasions, after he had sampled every article of food and eaten his fill of whatever appealed to him the most (how he loved it when there was sauerkraut), he would go to the lamp, make his rounds and settle down. In three or four minutes he would be sound asleep. Nothing disturbed him, not the clinking of silverware nor the clearing of the table between courses.

Robert's life was not all beer and skittles. There were a few crises, and then, especially if he was in pain, he wanted human companionship and comfort. We all had to be careful lest Robert were underfoot. Mildred always wears rather soft rubber-soled shoes, such as modified sneakers, in the house. One day, when I called for Mildred to take her to a luncheon, I found the household in anguish. When Mildred had gone into the bedroom to dress, Robert had been ensconced on the doctor's shoulder watching television. Usually we could hear the little feet tip-tapping across the floor, but this time he must have gone into the bedroom at a walk, which was not so audible as a run. Mildred had put on a pair of spike-heeled shoes. As she stepped back from the dressing table there was a scream—a scream of pain. She had stepped on one of the fragile little toes. She picked him up, the doctor came on the run and cuddled Robert close to his wool-shirted chest. How that toe bled! How so much blood could be in that tiny body is beyond comprehension. Mildred was in tears; there was blood all over Tommy's shirt. Some medicine to staunch the flow of blood was located and applied. (Where is the pressure point in a quail's toe?) The toenail end was gone. As animals so often do, Robert seemed to sense that he was being helped, and remained absolutely still while being treated, except for the plaintive little cries he made.

As we left for the luncheon, Mildred still a bit red-eyed, Tommy was walking up and down the living room with Robert practically under his chin, assuring him that everything would be all right, pretty soon it wouldn't hurt any more, as one would try to comfort a hurt child. The toe was still bleeding, though much less than at first. Of course, as soon as we got up from the luncheon table, Mildred telephoned home, and learned that the bleeding had stopped but that Robert would not yet leave the comfort of Tommy's hands. We arrived home from the luncheon about five o'clock to find Tommy still holding the little bird. He had held him all the afternoon. Robert held no resentment whatever and went willingly to Mildred, relieving Tommy. He hopped on one foot for the rest of that evening, limped badly for a day or two, then less and less. However, it was about five days before he walked normally. The toe did not grow out again—that is, the nail end or claw end—but a dark, very hard little nodule did form.

An almost incredible sequence to this accident occurred about three months later. One evening as the family were at the dinner table, and Robert had not yet joined them, they saw him fly up onto the coffee table and heard him give a shrill cry. He had miscalculated the distance and had struck that poor toe on the edge of the table, knocking off the little nodule at the end. He flew immediately to the dining-room table, went right to Tommy and with the same plaintive cries held up the injured foot. It bled, but only slightly, and was soon taken care of; but Tommy finished dinner with Robert in the crook of his arm. This time it healed much more quickly. An odd fact emerging from all this was that Robert continued to be fascinated by shoes.

Another time of worry was when Robert caught a cold. He picked it up from one of the grandchildren, who had a miserable cold himself. We always tried to keep Robert away from such contacts, but in this case it was impossible. He was really ill the next day. His eyes ran, his nose ran, he had to breathe through his mouth entirely, he was miserable. He must have had a temperature, for he drank quantities of water and sought out cool places, even going to the far end of the kitchen counter from the beloved lamp to lie on the formica surface. He did no grooming and looked dejected and ruffled up. He tried hard to clear his nose, working away at it with a claw. The doctor tried too, but cleaning a quail's nose is not easy. The breathing grew worse and he would not eat. Tommy considered an antibiotic, but dosages are determined by the body weight of the patient, and since his patient's body weight was just a few ounces, he was reluctant to risk any medication. He did give Robert a tiny drop of brandy. At the end of the second day of misery, we had determined to call in an excellent local veterinarian if Robert was not better the following morning. Fortunately, the morning saw great improvement and by the next day Robert was himself again, eating as though to make up for the many meals he had missed. He preened and he preened, stopping now and then to coo or chirp. Another crisis had been weathered.

All this increased our feeling of responsibility for him, our realization of his dependence upon human beings, and, if such a thing is possible, our affection for the tiny thing. This seemed to work both ways. Let a member of the family be temporarily couch-bound with some minor ailment, such as a headache or a sprained ankle, and Robert took over. He would spend hours at a time perched on the arm of the davenport near the pillow, seldom leaving the room and, when he did leave for a few moments, coming back to hop up, walk carefully up the patient's chest, to peer concernedly into his face, keeping his own talk muted and soft.

4

CHRISTMAS AND GRANDCHILDREN

CAPE COD favored us, as it so often does, with glorious pre-Christmas weather. One balmy sunny day after another gave Robert all the outdoor play he wanted. He ran through fallen leaves as any young thing does, and enjoyed above all else the woodpile. Over, under, through and around he went filling his crop with bugs of all sorts. No, that last statement is not accurate; he was quite selective as to bugs. The earwig is evidently as repulsive to bird as to man, and he left it strictly alone. Robert may have been a well-fed, protected house bird, but his instinct for finding food for himself remained acute. Spying a movement in the grass, imperceptible to us, he would run at top speed, make a three-point landing with feet and tail as brakes, and—pounce! Granddaddy longlegs seemed to be great delicacies. Making a beeline for the small center body part, he often had to swallow three or four times before the last wriggling leg disappeared. The process was one of concentrated enjoyment.

With his wonderfully protective coloring, it would have been almost impossible to locate him among leaves and twigs except for his almost constant chirps and trills. His plumage was by now very rich in tone, although the stripes down the sides of his face were still a soft coffee-with-cream color, instead of white as were those of his outdoor brothers. In fact, Tommy often wondered whether we should change his name to Roberta.

Several days before Christmas a crèche was set up on the buffet at the end of the dining room. The beautiful figures, hand carved by a dear friend, Mr. Herbert Plimpton, when he was in his eighties, stood before the stable, the whole scene backed with small pine-branch tips. So Robert had his introduction to the Christmas season. Since his sense of curiosity was so keen, we were not surprised when he spied it from the table and flew over to examine it. But we were surprised at his reaction. There was none of the excitement he always evinced when a bag of groceries was brought in; he just stood at the corner of the buffet and looked. Then very slowly he walked over to the crèche, looked at each figure carefully, then settled down in front of the shepherds, gazing at the Infant. And he stayed there. Mildred rushed for the camera, but she need not have been in any hurry, for the same picture was posed over and over. The end result was the Christmas card which both the Kienzles and I used a subsequent Christmas. Never once did he disturb a straw in the manger or touch a figure. Call it curiosity, call it fascination, call it what you will—he always seemed to fit into the Adoration.

The third day before Christmas, Robert walked out the door with Tommy after breakfast, started across the cement floor of the open porch and came to a dead stop with a little squawk of surprise. Out went the neck, and the investigation was on. On the floor lay a four-foot spruce tree, which Tommy had cut the previous evening. It took several minutes of stalking and investigating to convince Robert that everything was safe. How he enjoyed that tree! He played among the branches, neglecting even the woodpile. Every time he went out during the next three days, that tree was his objective. Here, his instinct seemed to fail him. Quail are definitely ground birds, but—perhaps because the spruce was lying down—Robert often nestled down on its upper side, sometimes almost going to sleep. (We often noticed that when he was apparently asleep on a lap, the eye toward the owner of the lap would be tight shut, but the one toward the room would keep watch.)

On Christmas Eve, the tree was taken in the house after Robert had gone to bed on his red velvet hat. This timing was not from any sentimental desire to surprise him the next morning, but because Mildred and Tommy knew that the decorating would proceed much more easily without Robert's help. They brought out the ornaments and went to work. Besides the grandchildren, who were coming the next day, Robert had to be kept in mind in the matter of decorations. No strands of tinsel were used, because, after all, they might possibly look too much like Robert's favorite sauerkraut. That precaution may have been unnecessary, for Robert was wiser than we sometimes realized, and almost never was fooled by appearance alone. I contributed one ornament. I had found a fallen branch to which a perfect bird's nest was firmly attached. Cutting off the branchlet which held the nest, I sprayed it with gilt and put inside it three small, fragile Christmas tree balls, one blue, one green and one red. We fastened the ornament near the top of the tree, which stood on an end table near the davenport. It looked lovely. Three little stockings were hung on the mantel—one for Thomas III, one for Kerry and one very wee one, really a doll's stocking, for Robert.

All was finished in time for the family to attend the midnight service, and when they returned, after a last look at the tree and the piles of packages, they finally went to bed. What fatigue is as delightful as that which comes to all families about one o'clock of a Christmas morning? The grandchildren would be there tomorrow, the beautiful scene in the living room would be a shambles in no time, and all would be as it should. After all, who should expect more than a few hours' sleep on this night of nights?

Breakfast was fairly early, since Tom and Nancy and the two children were to arrive midmorning, and then the wide-eyed ecstasy and the noisy excitement would begin, to continue throughout the holiday visit. This year would be the best yet, as a three-year-old girl and a five-year-old boy are masters of both ecstasy and excitement. But what about a six-month-old quail?

Robert did not get up until the family were at the breakfast table. He came into the room yawning as usual, hopped up on the table, had his orange juice and some toast and then—the dining room being a long ell off the living room—spied the tree. With a loud cry of surprise he flew over to the davenport with Tommy and Mildred following. He went slowly to the end where the tree was, giving the loud cries which sounded like "HUR-ry HUR-ry" with which he announced that the telephone was ringing, and then stood there looking, head first on one side, then on the other. Almost simultaneously he saw the stockings hanging on the mantel. Again he flew, this time up onto the mantel and, poking his head down to touch the little presents protruding from the top of each stocking, came at last to the tiny one belonging to him. A little sprig of chickweed was his treasure and he devoured it on the spot. He still seemed a bit bewildered and stood on the mantel looking over at the tree;

then to their astonishment he flew over to it, did not stop on the davenport, but landed right on the gilded nest. Even though it happened before their eyes, it was hard for Mildred and Tommy to believe. Had I not seen him there myself, later in the day, I should hardly have believed it. Robert did not know what a nest was, nor what it was for. Was it the bright gilt that attracted him? Maybe. But there were other ornaments quite as glittery. No one knows the answer but Robert. The indisputable fact remains that he settled down in the nest, thereby adding greatly to the Christmas morning loveliness. It was not just a momentary interest, as he returned to it often as long as the tree was up.

As Mildred and Tommy went back to finish their breakfast, Mildred remarked, "It almost seems as though that little bird tried to express thanks for all our work of last night."

"Yes," agreed Tommy, "and we will have it all over again when we watch the children. I never dreamed we would have this pleasure twice."

Robert had developed a technique during previous short visits from the children, who loved him dearly but loved him too much. His technique was to make a dash into the big bedroom and go under the big king-size bed to a spot in the geographical center of the sanctuary. That Christmas morning, that is exactly what he did. However, since he really loved human companionship, and since any excitement drew him like a magnet, he could not resist occasional forays into the middle of it all. How he loved the tissue paper, the ribbons and the boxes! However, he had no intention of being played with like a toy, and back he would go to safety. Let the grandchildren go outdoors to play with new treasures, and out came Robert to the living room and companionship.

So went the day, and so, finally, came night and bedtime.

The guest wing is at the opposite end of the house from his room, so he felt very safe when the small fry were finally calmed down and tucked in. That word "finally" was not used advisedly. The four tired grownups settled down for a game of bridge, which was just to Robert's liking. He hopped up on Tommy's shoulder, trilling and preening contentedly. The door of the wing opened: "I need a drink." The drink was provided—for both children, of course—once more they were tucked in. Soon the door opened again: "I forgot to kiss Grandpa goodnight." Grandpa was kissed, and again, quite forcefully this time, the tucking-in process was accomplished.

Once more they settled to their bridge game. Again they were conscious of the door opening, and *sotto-voce* among themselves agreed to ignore the small pajama-ed figure. It was Thomas III. He quietly and guiltily approached the table. Nothing was said; nobody noticed him. That is to say, no human being noticed him. Robert took matters into his own hands. Rising to his full height on Tommy's shoulder, he proceeded to read the riot act to the little boy. He scolded, he squawked, no sweet little trills now; and with his head bobbing violently with every squawk, he very clearly implied that if nobody else took care of this situation he would. And he did. Thomas III stared in bewilderment for a few seconds, then turned and meekly went back to bed, and this time it *was* final.

The grownups managed to keep straight faces through it all, and as soon as the door closed Nancy said, "Well! Let's take Robert home with us next week. He is a better disciplinarian than we are, evidently."

After a few hands of bridge had been played, Robert's curiosity about the fascinating disorder of the living room floor got the better of him, and down he went to investigate each toy, poke each movable gadget and enjoy the end of the day as he had its beginning, undisturbed and on his own.

When the family decided to have a last cup of coffee, they looked for Robert. Robert was not to be seen. Mildred called his name questioningly. As always, he answered. They finally located the sleepy little responses, and there he was, flattened out in the little gilded nest on the tree. It was such a picture that they left him there until their bedtime, then tenderly carried him to his red velvet hat.

Thus ended Robert's first Christmas.

5

IN THE SPRING—A SURPRISE

SPRING APPROACHED and with it a little gnawing fear in the hearts of Robert's human friends. Robert had come through the winter in fine spirits, if not in fine feather. In spite of having had few outdoor excursions and almost no really good dirt baths, he was as chipper and sweet as ever. With spring comes the molting season, and it certainly came. Feathers everywhere, first the little double-shafted breast and back feathers, and then more and more often a large wing feather. How a little bird of his size could shed enough feathers to fill a good-sized stationery box is a mystery. But he did. He looked bedraggled; for a time there was only one, rather ratty, tail-feather; but to our eyes he was still a delight. After all, families accept the awkward stages of their offspring with more or less equanimity and with no diminution of love and affection, and so it was with Robert.

The gnawing fear concerned all that was happening. Would Robert respond to the call of the season and the mating instinct?

We had long agreed that some end to this delightful episode was inevitable, and if it should take this form we would accept it as perhaps best for Robert. The outdoors quail began appearing in pairs rather than coveys, and we knew the season was upon us. As the weather permitted more time outside for him, we kept a careful eye out—just in case. But so far, he was not interested, in spite of hearing many a bobwhite call.

With the molting season we had expected his behavior to change with his plumage; we were prepared for him to be listless and mopey. However, this did not happen. Weeks of feather-dropping went by and Robert remained as gay as a cricket. We began thinking that possibly the good care and food he had had all through the winter accounted for his ebullient spirits. We should have knocked on wood.

One day suddenly Robert was very different. He did not get up until nearly noon, and when he did come out to the kitchen, he was not interested in orange juice or toast; not even in crisp lettuce. Robert did not feel well. As the day went on we became more concerned. Something was very wrong. He wanted to be held constantly. (His method of asking to be held was to back up to a human foot, squat down, and push with his tail till he was picked up and cuddled.) He wanted nothing to eat when lunch was served, and spent the time at the table nestled within the crook of Tommy's elbow, making sad little sounds, not of contentment but of distress. By now there was a general feeling of anxiety throughout the house, as there is when a child is not himself.

Robert stayed with Tommy until the dishes were done, then went to Mildred, who devoted herself to trying to comfort the poor little creature. When Tommy came into the house after working an hour or two in the rose garden (alone, as Robert did not want to go with him) Mildred said:

"I must get the roast ready for dinner. You take him for a while."

Remembering that a letter had to be written and ready for the evening mail, Tommy went to his desk, still holding Robert. He carefully put Robert down between his feet, forming an angle with heels together, which usually pleased Robert. He settled down, halfway resting on his side, and Tommy went on with the letter.

All of a sudden there was a scream. It can't be described any other way—it was a scream. Mildred came running in from the kitchen, and Tommy looked down in consternation, fearing that he had inadvertently moved his feet and hurt the tiny thing. Robert stood up, shook himself, gave a contented little chirp and walked off . . . leaving an egg!

Such excitement! I was telephoned to immediately, as were several other devoted friends. (To our surprise, the next issue of the local paper contained an article with a large headline: ROBERT SHRIEKS, LAYS FIRST EGG. There was by this time considerable local and even state-wide interest in Robert, so this was quite an event.)

Immediately after accomplishing this feat, Robert rushed to his (excuse me, I mean HER) tray and began eating as though she had never seen food before. And how she drank! As for the egg, she could not have cared less. She completely ignored it and seemed glad that the whole business was over. It was a full-sized egg, not a smaller, pullet-sized variety such as newly laying hens produce. It was the real thing. When we knew that Robert did not want to have anything to do with it, it was carefully placed in a little velvet-lined box to be admired by many callers.

After the first astonishment at seeing the egg on the floor at his feet, Tommy had worn a queer half-smile on his face. Now, as he stood looking down at the box on the coffee table, he said, "I'm not saying 'I told you so,' but I felt all along that only a female could have survived the vicissitudes that have befallen Robert. I guess we have the little girl we always wanted. I suspected this, and now the evidence is beautifully conclusive. Well done, Robert, well done."

Mildred had picked Robert up and stood murmuring comforting, congratulatory and female communications to her, adding, "Well, once again you have taught us something. It may take a while for us to learn to call you 'she' instead of 'he,' but we will, we will."

Then—the discussion as to her name. Must it now be Roberta? The question resolved itself. After a few halfhearted attempts to use the feminine equivalent, they gave up. Robert she had always been, Robert she was to her public, and Robert she would continue to be.

The ignorance of the general public on the processes of nature is appalling. So many people, on first seeing the egg, asked, "Will it hatch?" At first we took the question at its face value, explaining that it would not hatch because it was not a fertilized egg. Some covered their confusion, saying, "Oh, of course," but many others had to have the facts of life spelled out.

As would be expected, that question was always asked by children who saw it. The answer was very simple and always accepted. We merely said, "No, this one won't hatch. You see, there would be no father for the baby bird, and everything has to have a father."

Then came the next question: "What do we expect now?" We knew that there had been at least thirteen eggs in the nest where Robert was found, and checking in the encyclopedia and with ornithologists we learned that a quail's clutch was usually from twelve to fifteen eggs in number. We convinced ourselves that subsequent eggs would certainly come more easily, that Robert now knew what to expect and that we would have no more days of anguish such as that one had been. Should we fix a nest for her? After all, she had enjoyed the gilded nest on the Christmas tree. But she certainly had not hunted for any nest on this production day; she had wanted hands, human hands.

I recalled vividly having watched my mother when I was a little girl as she cleaned and prepared a chicken for roasting. I remembered that often there would be an egg just ready to be laid, with many smaller ones behind it in various stages of development. I tried to imagine Robert's small body cavity, filled with tiny embryonic eggs. An ornithologist who came to see her the following day suggested that we put the first egg in a nest-shaped place as a sort of decoy. He explained that often in the wild, when a marauding rat or snake destroyed some of her eggs, a quail hen would keep right on laying more until she had a full clutch.

After having lived with Robert for ten months, we should have realized that she would solve this in her own way, as she had solved the question of our giving her her freedom. She solved it: she just didn't present us with any more. That was it. Later on, the local paper announced, under the headline ROBERT'S A FAILURE. LAYS ONLY ONE EGG, that probably she had laid the one just to keep the franchise. Whatever her reason, she was through with all that foolishness. When people dropped in for tea during the days following the egg, the conversation almost always included speculation as to the why and wherefore of this unusual episode: On one occasion, as Robert was going from one person to another on the coffee table, hoping for bits of cake, she suddenly went to the center of the table, cocked her head on one side and stood motionless, surveying us. After what must have been a minute or two of this statue-like pose, one guest said, "If I ever saw 'no comment,' that is it."

Robert always had been, and continued to be, unpredictable. All the dire prophecies made about her had proved to be groundless. We recalled many of them with amusement: "That little bird won't live two days." "Wait till the first time she gets a chance to get out of the door." "The mating season will be the end of Robert's relationship with you." We should have learned to disregard such statements by this time, but another one could always make us wonder. The next worrisome prophecy we encountered came from a man who was an acknowledged authority on birdlife who had made quail his specialty. He had heard the whole story of Robert, and of course came to see for himself. He had to admit astonishment at some of the details, but maintained an air of wisdom on the whole subject. He finally made his pronouncement:

"Well, this can happen with birds in captivity. [But Robert was NOT in captivity.] They often survive the first mating period, or even the first migration period, in the case of migrating birds, which may be due to the fact that instincts have been sublimated. But this does not maintain through the second mating season. My advice would be to enjoy her while you have her but not to be surprised or disappointed next spring—if she stays that long."

In spite of his authoritative way of speaking, in spite of his reputation as a scientist and in spite of our customary desire to be courteous, we should have responded, "Oh, pooh!"

Since we now knew that she had reached full maturity, one frequent question bothered us, because we did not know the answer. The question was, "How much do you suppose she weighs?" But how were we to weigh her? We even considered taking her down to the post office and asking to have her weighed there. One day, an acquaintance of ours, Mrs. Nellie Barrington presented herself at the door and asked if she might see the quail. Mrs. Barrington was a very sweet and gentle person, who, in spite of her rather advanced age, was director of a very fine nursing home for elderly people. She had come quite a distance to see Robert, and was completely overwhelmed by her. She could not believe that she had an affectionate little quail in her hands or on her shoulder. While Robert was on her shoulder she asked if we knew how much she weighed. We had to admit that we did not, and explained our problem as to how to find out.

"Why, that's easy," said Mrs. Barrington. "I'll bring down the scales on which I weigh food for my diabetic patients. I'll bring it next week, on my day off."

She was as good as her word and appeared with a small platform type of scale. We put Robert on it and, obliging as usual, she stood perfectly still as the indicator moved up to . . . five ounces. We checked and rechecked. It was right; Robert weighed five ounces.

Mrs. Barrington telephoned me a few days later to say that she was having her seventy-ninth birthday soon, and to ask if I thought it possible for her to have her one wish granted on that day. She explained that the only thing in the world that she wanted was to have her picture taken with Robert on her shoulder. How nice it would be if all requests could be so easily granted, especially to the Mrs. Barringtons of the world. It was done and Mrs. Barrington was happy. How she loved the little creature, all five ounces of her.

6

A BIG PROBLEM

FOR ALMOST A YEAR I had lived in that idyllic condition of life that is generally conceded to belong to grandmothers. I had enjoyed Robert to my own satisfaction, I had played with her and escorted her on outdoor excursions, and I had glowed warmly at her evidence of affection for me. And I had had no responsibility for her health and general welfare. Now a problem appeared.

For years the Kienzles had dreamed and planned and saved travel folders for a European trip. It had been hovering in the background in the "someday" category. But all at once it crystallized. Their younger son, Don, was in Berlin, at the outset of his career in the State Department, and had fallen in love. His happiness could be made more complete only by having his parents there to meet his beautiful and wonderful Monika, to become acquainted with her family and to share in the engagement festivities. Don's eagerness to have them with him was matched by their own. The long-projected trip now had a real impetus. Their reaction vacillated between "Of course we will go" and "How can we; what about Robert?" To strangers this may sound a bit extreme, but Robert had become such an integral part of the family that consideration of her was of prime importance.

Several solutions were suggested. Put her in an aviary on a boarding basis? That was impossible at first glance. She had never been caged, and she had not only never been with other birds but she feared them. Take her to Europe with them? This was looked into but discarded as too hard on her, even apart from the quarantine laws in various countries. Friends and relatives were anxious about the outcome, as of course I was. I wanted their happiness and I wanted Robert's safety. Perhaps those two facts do not sound proportionate, but they were. I had a sneaking suspicion all along as to what the ultimate solution would be: I would take Robert for the three months of the trip. However, I said nothing, hoping against hope that something would "turn up." The Quakers have a delightful saying in times of indecision and stress: "Proceed as way opens."

At a tea one afternoon some people who knew the situation asked a neighbor what the Kienzles had decided about the quail. The answer opened the way. The neighbor said, "Well, I just don't know. They said the other day that it might be best just to give up the trip."

That did it. I knew I would have a guest. I spent a busy evening—busy though I did not leave my comfortable chair in the keeping room by the fire. I was making plans vigorously. Robert would have to be able to have her outdoor hunts and dirt baths. Robert simply could not go outside my house, even if I were with her, as I am surrounded by voraciously predatory cats, most of which were welcomed at my house at any time, and often fed. That would have to end. I went to the telephone and called Walter Peers, a very dear friend who was a builder.

"Walter," I told him, "I have a problem and I want an estimate on an enclosed patio in the angle of my keeping room and the main part of the house. Could you come up and see me about it?"

"You don't mean now, do you?"

"Yes, I mean now. I have made a decision, and if this is not settled now I may be tempted to change my mind, and I just can't change my mind."

"It is kind of late, but I'll come."

Together—Walter with his measuring tape and I with my plans—we measured, discussed one-inch or two-inch mesh chicken wire, dimensions, and so on, detail after detail. We came up with a decision on a ten-by-fifteen-foot "room" which had to be high enough to include the outside door of the keeping room. Walter figured, I waited, he gave me the estimate, and I telephoned the Kienzles.

"Would you trust Robert with me when you go away?"

"Would we! But do you mean it?"

"I do mean it, and all the details are settled here. Now you go to bed and dream of ocean liners and new daughters-in-law, and relax."

Their gratitude and relief were worth any amount of time, effort and expense. I went to bed rather smugly pleased with myself. Then the immensity of the responsibility I was assuming swept over me. What if the little bird were not contented here? What if . . . what if . . . what if? It must have been long after midnight when I suddenly recalled the one occasion when Robert had spent a night away from home.

On that occasion, the Kienzles had made plans to attend a

church conference, which was to be held not far from their older son's home in Lexington. Robert was then still sleeping in her carton, had ridden in the car several times and seemingly enjoyed it, and posed no problem for the visit. The plan was to be with the children in the afternoon, then leave for the night meeting, stay overnight at a motel, and attend the conference the following morning. That would give them plenty of time in the afternoon to see Robert ensconced in the son's home and all would be well. And all was well when they left her. They had taken her own food pan and water dish and she seemed content. What they did not realize was that her contentment was due to the fact that her family was with her and she felt secure.

Nothing untoward happened until it was time to put her to bed. The son put her in her carton, which was on a shelf not unlike its place at home. But to her it was not home. She kept getting out of bed, as she almost never did at home. Finally she was so exhausted that she stayed in, exhausted because she had gone all over the house, time after time, looking for Mildred and Tommy. Now, although she stayed in the carton, she cried constantly. Her cry at such a time would wring the hardest heart. It is a sad, pitiful and mournful sound, rather low and quite drawn out. Tom and Nancy reasoned that she would quiet down when they had gone to bed and the house was dark. Not so. She cried and she cried. Finally they took the carton into their room by their bed where she could know they were near and that she was not alone. It was a good idea, but it didn't work. The crying persisted. At last, in desperation so that they, at least, could get some sleep, they put carton and all into the guest room and closed the door. Nancy tiptoed to the door before morning and the poor little thing was still crying. Whether she had done so all night or resumed when she heard

Nancy—in spite of Nancy's trying to be so quiet—they will never know. But she was a very unhappy little quail.

Morning came at last and Robert was brought out to the dining room and offered a piece of lettuce, which she likes the first thing in the morning. She refused it. The children were got off to school, and even while they were having breakfast Robert was not interested. In fact, she didn't eat or drink all day. The crying stopped, but she just stood around dejectedly. It was a hard day for Nancy, who was alone with her. She put soft music on the record player, but it did not divert the lonesome little creature. Robert just stood on the arm of the davenport, waiting.

At about half-past two in the afternoon she suddenly gave a little cry, stretched her neck out, and Nancy saw the car turning in the driveway. (Robert always recognized their car, and later learned to recognize mine.) She chirped excitedly. The minute Tommy and Mildred came in the door, she flew to Mildred, cuddled down in the crook of her arm and almost immediately fell fast asleep. The departure for home was delayed to let the poor exhausted little bird make up her sleep. Finally, when Mildred's arm was almost paralyzed from having been held in one position so long, they woke her up. Tommy took her over where her food was, and she ate and drank greedily. She cuddled in Mildred's arms during the entire drive home, sleeping part of the time, but waking occasionally to emit a few little contented chirps. Once safely inside the house, like a child who has returned to familiar surroundings after an absence, Robert made the rounds of all her favorite haunts, and then, having satisfied herself that all was well again, hopped up onto her philodendron plant, closed her eyes, sang her sleepy song and indicated unmistakably that she was ready for bed. To bed she went, and there she slept soundly until after nine-thirty the next morning. All through that day she followed either Mildred or Tommy like a little shadow, showing them most convincingly that she did not like the idea of being separated from them. After that day all was back to normal.

The thought of this episode was not comforting me one bit, and it certainly was not soporific. I reasoned with myself that Robert was an intelligent little creature and that things would work out. However, I must admit that the "proceed as way opens" idea was not very convincing at that point. The situation brought to mind several things I had read. I recalled in particular a study by Dr. Konrad Lorenz, an eminent Austrian scientist, who reported a study of a male jackdaw. As I understand it, his theory involved a belief that a young bird taken into a human environment might adapt to such environment. The jackdaw of this case history attached itself to Dr. Lorenz from the very earliest stages of the experiment. This was carried to such an extent that when the mating season approached, the object of his courtship was none other than the doctor himself. The bird brought him choice delicacies, such as fat angleworms, as he would to a female of his choice. Finding the human being unreceptive to having the worms put into his mouth, he compromised by trying to insert them in his ear.

If we had ever doubted this scientific report, which we did not, the truth of it was forcefully brought to our attention. At the time of the first mating season, when we were wondering if she would leave us, the grandchildren were visiting. There were quail in the patio, and they knew full well that Robert was there. She gave no evidence at all of being interested in them, but for several days attached herself to five-year-old Thomas. Where he went, she went. If he sat on the floor playing with a toy, she was right beside him. They ate on the picnic table in the patio one day; during that time Robert was right beside Thomas. He could wrap her in a blanket and she would stay

with him quietly, though her tailfeathers did suffer a bit from being wrapped up. She wore herself out following him. It was a definite transfer of her feelings toward the little boy, and somewhat of a puzzle to him. One night as she followed him to his bedroom, he said, "Well, I guess there aren't many little boys followed everywhere by a quail." It was all over in a few days, but was a remarkable thing to witness.

Sally Carrighar, in her delightful book *Wild Heritage,* attributes to certain animals and birds the ability to make decisions, to feel and show affection, to play and to plan. She even goes so far as to suggest that some of them recognize property rights, fair play and so on. I myself have watched a covey of fourteen quail who frequent a neighbor's feeding grounds, and have seen them carry out a cooperative plan. One morning, after a very soft snowfall had deposited about four inches on the ground, one by one they came out of the woods, running very fast so as not to sink in the snow, to the place where they always found ground corn. Usually when they feed, as they always do in a close group, they head out from the circle so they can see danger should it approach from any direction. But not this time. They formed a very close little bunch, and then, all working together as in taking a dirt bath, threw the snow with wings, tails and heads, until it looked as though a small snow-blower were at work. In about three minutes they had dug a hole in the snow, with rather sheer sides, four or five inches deep. They had reached bare ground and were eating contentedly. After they had had their fill and departed, smaller birds came in to take advantage of the excavation.

Being convinced beyond doubt that Robert had a high Q.I.Q. (quail intelligence quotient), I asked myself sternly what good my years as a practicing child psychologist were if I couldn't work out something in relation to this particular individual.

Everything pointed to my being able to come up with some solution. In the first place, I was sure of complete and utter co-operation on the part of the Kienzles. In the second place, I was going to be working with an intelligent little creature. And in the third place, I loved her and had the strongest possible reasons for wanting to succeed.

As soon as I reached that point, I began to work on a definite plan. As always happens in sleepless nights, as soon as one stops worrying and stewing and begins to think constructively, the battle is won. The plan evolved so fast and so easily that I slept soundly through the 5:00 A.M. news broadcast.

7

FAIT ACCOMPLI

BY AUGUST 1, all was in readiness. The patio was built and equipped with delights for a quail. The soil in the geranium beds was loosened and softened, a large flat stump was installed, on which her V-8 juice could always be available, several clumps of sod were removed and replaced with clumps of chickweed, and lawn chairs were provided for my own convenience and that of any possible callers. How naïve I was in that respect! It was probably just as well that I did not know that in the three months Robert was to be with me, there were going to be about three hundred callers. Her guest book was brought over, of course, and by the time she left me there were names representing twenty-four states and two foreign countries. Writers, ornithologists, even a radio commentator from Boston, came to see and be conquered.

The date of departure of the Kienzles was set for Thursday, August 9. The time had come to put my plan into operation. Early Sunday afternoon Tommy and Mildred and Robert arrived, with Robert perched happily just back of the windshield on top of the instrument panel. Robert enjoyed riding in the car, whether it was the three-hour ride to Lexington or the shorter trips when she went on errands with Tommy and Mildred. Her favorite place was on the shelf in front of the rear window, where she would sit, looking at the scenery and often going from one side to the other as something caught her attention. This was all very well for her, but it proved to be so much of a traffic hazard that it had to be stopped. People in the car behind would see her in the window, and probably thought she was a toy. Then she would move and the rear car would come much too close for a better view. Many times a driver would come alongside the Kienzles' car and on a main highway this constituted a real danger. So most of her riding was done sitting in Mildred's hands, which she always liked, especially if Mildred would scratch her head.

They brought her in the house and we sat in the keeping room with the door from that to her patio open. At first she walked all around the room, neck outstretched, crest up, investigating everything. She examined the fireplace tools, looked the fire-screen over, hopped up on the drop-leaf table around which we were sitting and finally went to the door. She stood in the doorway, looking into her patio for about five minutes before venturing out. For all her curiosity, she is cautious. Finally she went out onto the brick step, pounced on an ant and then came back to fly up on the table again, as if to be reassured that her family were still there. After that she went into the patio for a good look around. She evidently found it to her liking, for she clucked excitedly when she discovered the chickweed and the geranium bed. We watched her as she stretched to her full height trying to see what was on top of the stump, which was about eight inches high and twelve inches in diameter; then she hopped up and took several sips of V-8. She was at my house three or four hours that time, busying herself while we three human beings talked about the coming trip and made a list of just what was to be brought with her in the way of equipment.

On Tuesday she was brought over again, and this time the Kienzles left immediately, as planned, to do some last-minute shopping; they were to return for her about five o'clock. This time the door to her patio happened to be closed. I, of course, watched closely to see what she would do. After a very cursory look around the room, she went to the door and stood there looking up at the doorknob, making little interrogatory chirps. I let her out, but within two or three minutes she was in the room again standing by me. She kept repeating this until I finally realized that she wanted company out there, so I took my book and joined her. She was as happy as a clam at high water. However, when I went into the house to answer the telephone or for any other reason, she was right at my heels. But she did eat and she did drink, and she even took a little nap on the back of the other lawn chair in the sun. I really relaxed then, and felt that all was going to work out well. She was quite excited when the Kienzles came to get her, and greeted them with the very distinctive and rather loud cries with which she greets almost everybody. My report of the afternoon delighted them and off they went, not to return again until Thursday morning on their way to the airport.

Robert had almost as much luggage as the travelers. Her equipment included such things as her philodendron plant in its familiar milk-glass bowl, her feeding tray, a large bag of wild-bird seed and another of sunflower seeds, her carton with the red velvet hat, one of two wiggly toys which she sometimes played with, the guest book of course, the boudoir lamp which was to spend the time on my kitchen counter to make the place seem as much like home as possible, and a pair of Tommy's gardening shoes. These they brought because often when they returned from shopping, they would find Robert cuddled down between the toes of these same shoes. I must admit that they added little to the decor of my keeping room, but who cares for decor when a guest's happiness is being considered!

As the time approached for them to leave, I witnessed one of the most poignantly touching scenes I have ever seen. To my mind, there is something especially appealing about a big, athletically built man who loves any tiny animal or a tiny baby. After Mildred had held her lovingly and kissed her good-by, Tommy picked her up and, holding her cradled in his hands up under his chin, walked away into the kitchen, where we could hear him assuring her that he would be back, that she was going to have a nice time and that he wasn't abandoning his little girl. I did not know him when he was a practicing physician, but I couldn't help thinking what a doctor he must have been. Finally he handed her to me and, after a sort of second-thought good-by to me, they left. I think many grandmothers who have taken on the care of an infant grandchild can realize my feelings as I watched them drive out of the yard. It was here. The responsibility was mine; I was in for it.

I busied myself getting Robert's things settled while she, to my surprise, investigated the rest of the downstairs part of the house.

The red velvet hat was put in the carton up on a high shelf in the little bedroom off the kitchen which in these old New England homes is called the "borning room." I had moved from an upstairs bedroom to the big downstairs bedroom in order to be nearer her. I put her feeding tray with its seeds and gravel near the fireplace, as it was at her own home, and put her water dish on a newspaper in the kitchen, for a calculated reason. I was consciously keeping myself busy.

She got on the table with me while I had my dinner, took bits of lettuce from my salad, took a few sips of coffee when it had cooled, called me in a loud, shrill voice whenever the telephone rang and seemed in every way at home in her new surroundings. Still I knew that I was going to feel a lot better once I had put her to bed successfully. Immediately after dinner she got on the edge of the sink while I washed the dishes, and then hopped up into the philodendron plant, stood on one leg and closed her eyes. I got her glass of orange juice ready which she always had at home after being put in the hat, and after pulling the shade down in her bedroom window I took her in, stood on a stool and gently put her to bed and gave her the juice. I rubbed her head a few minutes—something she loved—she made her little sleepy sounds and I softly left the room and closed the door. I was very quiet and even stifled a sneeze. All was still for about five minutes. Then I heard a thump. I rushed in, fearing she had fallen out of bed in the strangeness of her surroundings. Not at all. She had pushed the velvet hat out of the box onto the floor. The open end of the box faced the room, so this was easy for her to do. She was sitting on the windowsill and spoke to me the minute I opened the door. I picked the hat up, picked her up and established her once more for the night—I thought. Again I tiptoed out and closed the door. Again in about five minutes, the thump. Again the hat was on the floor, but this time she was on the bureau between two large soft balls of mohair yarn. After this had happened three times, and after I had again found her on the bureau, I let her have her way. I got a soft facecloth, put that on the bottom of the box, put the two balls of yarn in with Robert between them, talked to her softly for a minute or two, telling her to be a good girl and go to sleep, tiptoed out, closed the door and heard nothing more till about eight o'clock in the morning. From that time on, she would have nothing whatever to do with that hat. In fact, I put it on the bed in that room, where it stayed for a day or two until she got up on the bed herself, pushed it off onto the floor and then pushed it until it was almost under the bed. The hat was all right at home, but not here. I removed it entirely and there never was any more trouble.

There were to be other changes. In her own home she often slept until nine or ten o'clock, even though she usually went to bed about seven. Not here. She was up the minute she heard me stir. She always called to me and I went in and took her down, carrying her over to her water dish on the newspaper. She always drank the first thing in the morning, made the two deposits and then was ready for the day. It took four mornings for her to learn to go to the water and the newspaper, and to stay there till everything was accomplished. During that first day I remembered that I had not given her the half-teaspoonful of cooked rice which she always had after she was in bed. I had nothing in the house but some cherished—and very expensive—wild rice, but I cooked a little of that for her the second night, and she loved it. Indeed, she pecked at the spoon after she had finished it till I went back to the kitchen for a little more. From that time on, she never touched white rice. She was a very discriminating young lady.

I suppose it would be ridiculous to impute to Robert any real plan to discard reminders of home, so I shan't do it. But the fact remains that, far from retreating to the garden shoes for comfort, she avoided them like the plague. Whenever she had to pass them she went the long way around with her neck stretched out to one side, eying them with suspicion. I was not sorry to realize finally that she was not going to associate with them at all and to put them away for the duration.

I had one worry, which, like so many worries we have, proved groundless. One half of my large kitchen range consists of four burners which are fed from an outside oil tank. That part is always hot, except in the summer when I turn them off and use the gas-burner half. The oil burners ensure a warm kitchen and a steady, low oven heat, enabling me to have such long-baking delicacies as Indian pudding, baked beans, and so forth. But the hot surface did cause me some concern. I had a cover made of mesh chicken wire; it stood about an inch above the stove surface. While the wire was quite warm, it could not have burned Robert. She frequented all the other work surfaces in the kitchen—the sink and counter and cabinets—but not once did she go near the stove. If I left the oven door open for any length of time, she would scurry over to the mat in front of the stove and squat down to enjoy the extra warmth. In her own very modern home the electric burners in the kitchen are flush with the long counter on which she originally was hatched. She was free to go wherever she chose there, but when she passed the burner part she always scooted along the cool outer rim of the counter. This was true whether or not the burners were hot. Can there be any question as to her intelligence? Over a year later, when she spent Christmas with me, at which time the furnace was on, I covered the hot floor register with wire. I need not have. She always went carefully around it even when running from room to room.

While most if not all my worries proved to have been needless, by the same token one thing on which I had relied did not work out exactly as I had planned: her patio. She always loved it and would beg to go out there when the door was shut, but she would not stay out there alone. Whether this went back to the fact that when she was outdoors at home at least one person always accompanied her, I do not know. Because of gnats and other little flying things, it was necessary to keep the aluminum

Epicure
WILD
RICE

screen door closed. Friends—no, not friends: acquaintances—dismissed this difficulty with the remark, "Just put her out there and leave her there."

I tried it. I tried it several times. When out there by herself, she paid no attention to a choice ant, she did not go into the geranium bed—she just stood on the brick step and cried. I tried shutting the inside door so that I could not hear her. But at those times I stood by the kitchen sink looking out the window where I could watch her. There she stood, looking up at the door, and I could tell that she was crying by the regular motion of her throat. I timed my last such attempt, and after twenty minutes she was still standing there on the step looking so pitiful that I went out to her, picked her up and cuddled her, to her intense relief, saying, "I give up. You win." After all, what is a mere three months?

I had made arrangements with a bird-loving neighbor Helen Lindorff to baby-sit when I had to be away longer than it takes just to do some marketing. This worked out beautifully, to the complete satisfaction of all three of us. On the occasions when Robert was alone for perhaps half or three-quarters of an hour, I wondered where she would choose to wait, since she had rejected the shoes. She solved this the third day. From that day on, she was always to be found on top of an old Bible on the blanket chest in the front bedroom. It had two advantages: she could look out the window and it was near a large table lamp, which I often turned on for her if it was cool in the bedroom. She even went so far as to get up there *before* I left, when she saw me put a coat on and take my car keys. The only other place I ever found her, which was always when I had to leave her for a few minutes in the evening before she went to bed, was in the living room on a New Testament which was on a little table by a big chair. I know as well as the next person that this was sheer coincidence, but it was an odd coincidence. The odd part of it was that the two volumes were not alike at all as to binding. The Bible was a soft leather; the other book had a carved olive-wood cover. But the indisputable fact remains that the Bible and the Testament were her refuge in times of lonesomeness.

All this did not constitute the inconvenience it might seem to, because I had all my evenings free. She always went to bed between six-thirty and seven; and after her wild rice and V-8 and a few minutes of a sleepy little song, she never stirred until I got up in the morning. My friends were most cooperative. Any invitation to a tea or a party was tendered with the opening question, "Can you get your baby sitter for Thursday afternoon?"

Thus began one of the happiest, and certainly the most instructive ninety-six days of my life. What a lot I was to learn.

8

I LEARN

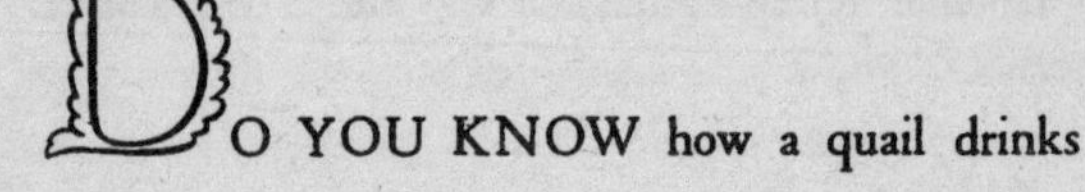

O YOU KNOW how a quail drinks dew? I do.

Do you know the unvarying technique of a dirt bath? I do.

Do you know what sort of insect a quail rejects and what kind of seed a quail cannot manage? I do.

Robert's refusal to stay out in the patio without me proved a great boon. Not only did I get a great deal of reading and hand-sewing done, but my eyes were continually being opened to facts I did not know and to which I had never given a thought.

On several mornings she and I were ensconced while the grass was still wet with heavy dew. Since it was impossible to have the grass mowed inside the enclosure, it had grown to a length of five or six inches within a very few weeks. All of us have seen grass sparkling with dew. The poets have written about it and singers have sung about it. But I wonder how many have really looked at those sparkles? I found that the drops usually form on the blade of grass where it has started to bend over. Because of gravity, I suppose, there is almost always a drop or two further along toward the tip of the blade. I had never known or noticed that a blade of grass is grooved, amounting almost to a lengthwise center fold. Robert always walked around seeking out the most productive blades; then, starting at the lower, pointed end of the blade, she opened her bill and ran the lower mandible up its entire length, scooping up the drops as she went. Then she would stand back and swallow. If she had enough water to make an especially big swallow, a tiny droplet was often left on the tip of her bill. This was always shaken off and I have seen those droplets fly through the air a distance of three or four feet. She drank much more water that way than she did from her own dish, and I wish I knew whether it tasted better than tap water or whether she had inside knowledge that she had better make time while it was available, since we were always out there on sunny mornings and the dew was gone very quickly. Morning after morning, I watched, often down on my knees with my head a matter of inches from hers so that I could observe the process. I suppose all birds do this, but I had not known it.

I have seen that she showed real discrimination as to insects. Ants, flies and occasional mosquitoes were her preferences. If I moved the stump, exposing grubs underneath, she would start toward them, but would stop about three feet from them, knowing they were not good to eat. She would get away from a wasp as fast as possible, but I watched her catch two honeybees. In each case the bee was on a geranium blossom, and Robert attacked it tail-end first. In each case she ate the soft body part, leaving the rest.

If I saw a spider before she did, a few quick taps on the ground brought her running at full speed. Sometimes I would see a little spider high on a wall or on a ceiling. Since the ceilings in these old Cape Cod houses are low, I could always hold her up and she would get it in one quick flash. A friend once brought her an angleworm, which I thought she would like. But she would have none of it. I tried her on angleworms several times, to no avail. However, she did relish small inchworms.

She was equally discriminating about her bath. I did not know that quail never bathe in water. They want dirt. One of Robert's admirers thought she ought to have a sandbox, and went to a great deal of trouble to get sand. Someone told her that it should not be seashore sand, because of the salt. So she tried without success to wash and rewash some sand till it was free of salt. It didn't work. One day she saw a lovely pile of clean sand where some construction work was going on, and begged a shoeboxful of it. We fixed it up in a flat wooden box, and it looked most enticing—to me. Robert, always curious about anything new, walked into it, scratched a time or two and then walked off, completely uninterested. She wanted dirt. And she wanted dirt which was slightly damp. Her choice was dirt in the geranium bed, soft on top and a bit damp as she worked down into it.

The bath technique never varied. With just her feet, she first loosened up the soil until she had a little hollow. Then, with the loose dirt around her, she went to work to throw it all over herself. She would work first with her head, which she put down so that the side of her face was in the dirt, throwing it back and over her, first on one side and then on the other. When she was in a little deeper, the wing work began. How she could make it fly with those wings, working from one side and the other. In a short while she would have a hole large enough to accommodate almost her entire body; also, by this time she was beginning to get tired. She then would lie on her side, making little convulsive, jerky motions, and the last stage of the process was to push with both feet against the side of the hole, seeming to be rubbing her back against the opposite side. Then came the most amazing part of it all. If I had not seen it before seeing it at my home, I would have been terrified. But after she had had enough, she would stand up and actually stagger out onto the grass. She looked and acted like a drunken person; she would fall from side to side, get up, fall again, until she regained her equilibrium or perhaps until she was rested. Then—watch out! Then came the shaking. I learned to wait until she had shaken three times before disturbing her or taking her indoors. I made a big mistake one day when a sudden shower came up while she was in the staggering stage: I hurriedly took her in and put her on the drop-leaf table, both of whose leaves were open at the time, fortunately. Well! Such a mess as I had! She shook, and shook twice more. Then I went to work. I brushed the dirt into a little heap on the table, and by actual measurement I took up two tablespoonsful of it. How much more there was on the floor, on the windowsill and on me, I do not know. But there was a lot. It was my own fault, so I cleaned it up uncomplainingly.

Tommy had told me that after a bath, especially if it were the least bit cool, she wanted to be held. This I had to learn for myself, as I had been inclined to discount it a bit. She made her desire known at such times by backing up and pushing against my ankle, talking all the time until I picked her up. She would cuddle in my lap, with both my hands around her so that just her head showed, and there she would stay for about ten minutes. She even shivered a little at first. Once warm and rested, she was off about her business.

Her fear of birds never failed to arouse my sympathy. If any bird, even a little chickadee, flew over the patio, no matter how high above the top it might be, she gave a little cry, fluffed up and flattened herself against the grass. If she were near enough to a geranium to dart under it, she did so. She never got over this fear. When I was studying advanced psychology in college and university, so-called "experts" were casting doubts on instinct. If I had accepted the idea then—which I did not—I certainly would have rejected it now. What else could this fear have been but the instinctive fear of the hawk? I understand that mother quails have been seen and heard warning their brood that a hawk is overhead. At such time the babies scurry into underbrush and the mother flattens and "freezes." But Robert never had such parental instruction.

What would you call it, if not "instinct"? And what told her that an earwig, ugly, repulsive-looking insects as they are, must not be eaten? Something told her. What made her turn away from her little glass of orange juice, without even dipping her bill in it, when it was the least bit soured? When she did this I often picked up the glass and smelled it. There was no odor as far as I could detect. If I tasted it, I always found that it was a tiny bit rancid. When offered fresh juice, she drank lustily. One morning I forgot to replace the little glass with a fresh one. She went to it several times, until finally I caught on and got a fresh one. At once she took sixteen swallows as though she could never get enough. She did not take a mouthful then hold her head back to let it run down, as a chicken does. She took perhaps three dips, waited a second or two, then swallowed audibly.

Had she been larger and a bit ferocious-looking, she would have made a good watchdog. I could be engrossed in a book and know at once from her really loud calls that a car had turned in my driveway.

Except for the flying birds overhead, she seemed to feel completely secure in her enclosure. One day when I was sitting outside with her, buried in a book, my back to the wire because of the sun, I became aware that she was clucking in a tone rather like that she used when she had found a good spider, but with

more to it. I turned, and to my horror there she was, walking back and forth along her side of the wire, while safely on the other side sat not one, not two, but three CATS. One was flattened out with tail twitching, but the others were just sitting with their noses almost touching the wire. I saw at once that there Robert was in no danger. One cat even put a paw tentatively through the mesh of the wire, disturbing Robert not at all. She had no instinctive fear of four-legged animals.

As a matter of fact, she made friends with a dear little white French poodle who used to come almost every day with his mistress. I knew him well and he knew me and the house. We kept him on the leash all the time for the first week or so, but soon found out that Robert was not afraid of him and that he was merely interested in her because she moved. Occasionally she would snuggle up beside him and stay there until he reached around and gently nudged her, wanting her to move.

Toward the end of her stay with me, Robert developed an amusing game with the poodle. By that time he had lost all interest in her, moving or not, and since he was an old dog he would settle down and go to sleep. He was most often in a spatchcock position, front legs stretched out before him and hind legs stretched out behind, lying on his stomach. By now she had assumed the role of the aggressor. Stalking around in back of him, with neck outstretched, she would get close to his rear end, select one white curl of his hair, then like a flash grab it and give it a yank. That always galvanized Jackie, who leapt into the air, while Robert would retire delightedly, only to repeat the performance later. I wish I had more and better pictures of the two of them together, but as any photographer will tell you, it is very difficult to get the animal pictures you want. Later, Robert developed into a regular ham before a camera, apparently liking even flashbulbs.

One event in her relations with animals is so incredible that I almost hesitate to record it. All I can say is that it is true in every detail. She was out one day with Tommy and Mildred while the doctor was weeding in their patio garden. Suddenly a little rabbit, perhaps one-fourth grown, hopped into view. Robert took off after him with the Kienzles following as best they could. She chased that poor little animal until it was so exhausted that it just collapsed, at which point Robert settled down right up against its warm fur. Finally Mildred tapped on the ground and she came running to see what tidbit might be forthcoming, and was taken back into the house.

Her fear of birds did not extend to birds on the ground. Once in a while even when I was sitting quietly in the patio, a little sparrow would come in through the wire to get some of the wild-bird seed that was always there. At such times Robert was not at all disturbed and I have seen the two of them eating contentedly.

9

CALLERS

CALLERS! As I said earlier, it certainly is fortunate that I had no forewarning of the people who would come to my house. I was blissfully ignorant on that August 9. Never, as long as I remember anything, will I forget our first Sunday together. Word had spread that Robert was with me. On that Sunday we had nineteen callers. Once there were seven people at once. Not once did Robert let me down. We all want children to appear at their best before company, and we all know they don't always do it. But Robert cooperated unfailingly.

She couldn't manage sunflower seeds: I understand no quail can. She was passionately fond of them if they were cracked for her and the meat taken out. She would almost dance in front of me when I was cracking them for her. I always kept a few on hand all picked out for her. Everyone who came was entranced with her and, of course, each one wanted to feed her. I gave each person a sunflower meat or two, showed them how to tap on the floor to see her come running, and how to hold it a bit above her head to make her stretch up to get it. She was at her most beguiling, but I learned that day that she could have too many sunflower seeds. I discovered later that eight or ten of them constituted her limit. Beyond that she was distressed. She showed this by standing still, stretching her head back and out, back and out, then opening her mouth very wide, wider than would be thought possible. This seemed to relieve her. There was no sound to it, but it must have been sort of a refined, silent belch. I had overloaded her poor little insides with these rich bits. Where was her own protective instinct? I don't know. If any amount of the meats were left in a dish she would never overeat; but when one person after another was offering them, she just cooperated too well.

The last guest left about five-thirty. I had had my Sunday dinner at noon, and was too tired to do more than get a cup of tea. Robert, as usual, was on the table as I relaxed with my tea. All at once I noticed that she was standing listlessly on one leg, her eyes closed, her head drooping till it was almost touching the table. Robert was just tired out. I picked her up, held her against me for a minute or two, talking comfortingly to her, and took her to her room. Without benefit of either orange juice or wild rice, I put her between the balls of yarn, pulled the shade, shut the door and I believe she was asleep before the latch clicked. She awoke even later than usual the next morning.

I was to realize later how many little cat naps (perhaps that is not the right word for a quail nap) she took during the day. She had had no naps all that long Sunday. The result was a completely exhausted little bird.

I also learned the following week that she greatly enjoyed an afternoon nap with me. These naps developed into quite a routine. I had cut up an old woolen blanket to make smaller ones for her. I kept them in her favorite places, such as on the old

Bible, because if she was there for any length of time there were bound to be droppings. The blankets were easy to wash and saved a great deal of work.

The routine was for me to get my book, lie down on the day bed in the keeping room, spread the blanket on my middle and call Robert. She knew exactly what was happening and was always impatient to hop up with me. Her little sleepy sounds at these times were captivating. I am sorry I have no recording of them, though the Kienzles have a tape recording of most of her other calls. It took her a few minutes to get completely settled, but it always ended with her lying on her side, with her feet stretched out and her head stretched back in complete relaxation. I could turn or readjust my position and she slept on. The only thing that disturbed her was the telephone. When it rang she would awaken from a sound sleep, fluff up and literally yell for someone to answer it. (I do have proof of this in a photograph.) Once both of us slept uninterruptedly for almost two hours. I really believe that she looked forward to these naps, for the minute I got a book and the blanket she paced back and forth by the day bed. The truth is that we both enjoyed those naps. I was always surprised by the amount of warmth that radiated from that tiny body.

But to get back to the callers. Some came from curiosity, others from a real interest in birds; some came doubting the veracity of the stories they had heard about Robert; some came from a scientific interest in the quail; and many came not really in disbelief but wanting to see for themselves that stories they had read in local papers were not fictional. Those who came with disbelief shed it visibly, and the remark oftenest heard—not distinguished for its originality—was, "Well I never would have believed it if I hadn't seen it with my own eyes."

By far the majority of those who came experienced the wonder of it all. It was unprecedented, unheard of, to have a quail, the shyest of birds, run up to greet them vociferously and so eagerly. I was forced to keep my house in order, because I had purposely had the patio built with no door opening from it to the yard. Therefore that entrance to the house was cut off—which meant that everybody had to go through my kitchen. It was probably good for me. I need not have worried or apologized if there were dishes in the sink, since people had eyes for nothing but the beautiful, friendly little bird. Who could resist not only her greeting but her blandishments as she hopped up on a strange shoulder and cuddled under an ear?

One caller was an ornithologist from Canada who had read about the one-egg episode. It was hard to make him believe that Robert had laid just that one and no more. Was I sure she hadn't hidden others? Was I sure she didn't lay more outdoors? He finally believed me, but begrudgingly.

For the sake of truthfulness, I must admit that there was one person, and one only, among her hundreds of visitors, who resisted her charm. This was an artist who came from Provincetown with her sixteen-year-old son. I noticed that the young lad had two or three cameras with him, but that was the rule rather than the exception among people who came. They stood outside the enclosure watching Robert, who remained on my shoulder. Finally I asked them to come in. The artist stayed outside, but at once her son was in the patio, camera poised, waiting for just the right shot. He took several pictures, and our local camera shop was so impressed with one of them that they asked permission to put it on a commercial postcard. I explained that, as far as I was concerned, they could, but that I would have to get clearance from her parents, who were now in Europe. An airmail letter brought a response within a week, granting their permission. The card has been sold by the thousands. It was odd that out of the visit of the one and only person who did not capitulate to Robert, came some of her most dramatic publicity. At least, up to then it was her most dramatic publicity—but little did we know what lay ahead. In all fairness to the artist, I will say that in subsequent visits she almost learned to love the quail.

Robert was sensitive to color, and here again had distinct preferences. High on the list was red. She always selected the red candies from a dish, chose hearts and diamonds when joining in a Canasta game, and picked out scraps of anything red when I was sewing. I was interested to read that only monkeys and birds have both rods and cones in their eyes, as does man, thus enabling them to distinguish color. I knew she could tell color but was glad to have this knowledge authenticated.

Often there would be three or four friends here in the afternoon and I would serve tea. Robert loved these occasions, and was always a member of the party. She would fly up onto the tea table and, clucking delightedly, would go from one to another guest for a nibble of cookie or cake. She had very cosmopolitan tastes in food and would try almost anything. The one exception was banana. She always walked away from bananas, to the extent of ignoring my breakfast cereal if there were slices of banana on it. Otherwise she liked cereal, and when there were fresh berries on it, any kind of berries, her joy knew no bounds.

Most people called ahead to make an appointment to see Robert. However, some, even strangers, just appeared at the door. When I greeted such a stranger and heard, "Pardon me, but is this the place where the little—?" I knew. They usually came in a bit apologetically, saying they hoped they were not inconveniencing me, that they would stay only a minute, and so on. Those minutes were always magnified, often reaching hours.

One such protracted minute involved a young man who

makes and sells beautifully carved birds. He got word to me that he was going to some sort of gathering and would like to stop for a few minutes if he could see the quail. I was more than glad to have him come, and immediately thought what a lovely Christmas gift for the Kienzles a MacKenzie quail would be. He came, accompanied by his brother. Robert greeted them at the kitchen door, investigating their shoes as she always did. Both young men stood motionless, almost speechless. I suggested that we go into the other room, and the bird-carver said, "May I look at her a little more?"

"Oh certainly," I replied. "She will come with us."

They came into the keeping room very cautiously, as though they were afraid they would frighten her away. They told me, after an hour or two, that they were astonished that they could actually see and touch her. They had expected a fleeting glimpse of a little shy wild thing as she ducked under some piece of furniture. We all sat around the table with Robert on it, and I leaned back and listened. I finally had to get them paper and pencil. Those boys knew their bird anatomy. They conversed between themselves about the angle of the lateral this and the dorsal that, examined a feather Robert dropped as though it were a jewel. Once the carver remarked, more to his brother than to me:

"I've carved lots of quails, but I always made a big mistake, I see now. I never realized that there was a distinct dividing line down the breast center. Look at that! She is really double-breasted! And do you see the iridescence on the gray back feathers? They are almost blue in tone. This is wonderful."

It turned out that he had used some little dead bodies as models, and also some nicely mounted ones. But this was Life. After they had been with me two hours, one of them looked at his watch and said in amazement, "Do you know what time it is? It is almost five!" and off they dashed, thanking me almost reverently for the afternoon. I wonder what happened to the hostess who had been expecting them those two hours.

I wish I had kept an actual count, but I feel quite sure that well over 50 per cent of her callers left asking if they could bring So-and-so, who would simply lose her heart—or So-and-so, who was so especially fond of quails, and so on. Of course I always said they might. And they did.

One very charming group, six strong, made arrangements to come from a neighboring town, bringing two house guests with them. They announced that they believed they were relatives of Robert's. I was accustomed to having people identify themselves with her by such remarks as, "I think she likes me," so I took this in my stride. I politely asked the reason for the statement and one woman said, "Well, my mother's name was Woodcock; my grandmother's name was Quail, spelled Q-u-a-y-l-e, I believe; and my grandmother's name was Martin." I agreed readily that they certainly all belonged to the same family.

One chance encounter developed into a real friendship. In the local bakery one day I met a friend and showed her a lovely picture of Robert which I had just picked up at the camera shop. A stranger, who I found out later was touring from Illinois, was in the shop asking the way to a charming little church which happens to be on my road. The saleslady asked to see the picture and, as a matter of courtesy, I showed it also to the stranger. Such a response! She asked all sorts of questions and seemed to have lost all interest in the church. I told her that if she followed me I could show her the way to the Church of the Holy Spirit, and asked if she would care to stop at my house and meet Robert. She would indeed. She followed me, and she and her husband got out laden, in true tourist fashion, with cameras.

Mr. and Mrs. Pope were delightful people and that Christmas I received an attractive card made up of pictures of their travels. Prominently among them was one of Mrs. Pope, looking like a delighted child, with Robert on her shoulder. It developed that Mrs. Pope is a columnist for a large Midwestern newspaper and has adequately taken care of Robert's publicity in that area of the United States. A letter from her asked for some pictures of my home and said, "I'm writing up the Cape Cod trip as sort of a travel feature, but to tell the truth, all else seems to pale in comparison to Robert and our visit in your home."

And so it went. It was a summer rewarding for me in human relationships. I glow when I think of the impact these five ounces of bird life made, and how many lives she enriched, to say nothing of her fame.

10

JUST THE TWO OF US

IT IS OFTEN SAID that you cannot really know a person until you live with him, or travel with him. I am sure this applies to animals as well as to human beings, and I am doubly sure that it is true of quails.

Much of her charm and many of the enchanting things she did were reserved for times when we were by ourselves. I would expect this to be true, as indeed it was, of her periods of showing affection, getting on my shoulder and rubbing against my neck with little coos of endearment; but I was a little surprised by "the game." She usually entertained herself well, or busied herself, when I was working. But at times she wanted to be amused. This most often happened over a bit of lettuce. Instead of eating it, she would come running to me, holding the lettuce in her beak. Then, pushing against my ankle with her tail and looking up at me with a real gleam in her eye, she let me know that she wanted to play "Chase Me."

I usually stopped what I was doing and, bending over her, saying, "I'll get you! I'll get you," would start the chase. Off she would run, with me after her, across the kitchen, through the living room, across the hall and through the bedroom back to the kitchen. Round and round we went, sometimes for as long as five minutes. Sometimes I doubled back on her and met her face to face in the hall, at which point she would brake to a stop, pivot in a flash and reverse the path. It was definitely a game; if she had been trying to get away from me, all she had to do was disappear under a table or a bed. The chase continued until she was tired, at which time—*gulp-gulp*, the lettuce was swallowed and the game was over. If she chose to play while we were at the table, she would run around the surface, in and out of spaces between sugar and cream, around vegetable dishes, but never leaving the table.

I often wished that visitors might see this. However, I think the reason they did not is plain: Robert wanted to play when she was a bit bored. When guests were there, she was so engrossed with them, or perhaps so satisfied with the attention she was receiving, that nothing else was necessary.

Once the game resulted in an anguished, sleepless night for me. I had been sewing, putting a binding on a baby bib. I had snipped off a two-inch piece of bias binding, and it was red. As quick as a wink, Robert was off with it, with me after her. I can still see that outstretched head with the banner of red tape flying in the breeze. Whether because the tape was heavier than a bit of lettuce, or for some other reason, the game was abruptly abandoned. Within a minute, to my consternation, she stopped —and with about three bobs of her head she swallowed the tape. This happened after supper, just before her bedtime. I gave her a little less wild rice than usual that night, all the V-8 juice she wanted—and worried. That was really a bad night for me. Tommy was an ocean away, and there was no one I could turn to for professional advice about a quail.

The next morning I went through the two big droppings with a toothpick—nothing. I watched her carefully that day, telling two callers what had happened. They must have been concerned too—or at least interested, and sorry for me—because shortly after they had left, I received a telephone call from a very elderly lady whom I did not know. She said she had heard about the bias binding, and told me just what could be done. It did *not* comfort me. She had been brought up on a farm, and said she had often held hens who were suffering from "crop bind." There was nothing to it. She had watched her mother slit the bulging crop with a razor, remove the obstruction, which might be a bit of fine wire, a pebble or even a packed wad of feathers. Hair balls I had heard of, but not feather balls. She tried to assure me that it did not seem to be painful to the hen, and that after staying in a box for a few hours it was always put out with the rest of the chickens none the worse for wear.

My consternation must have been evident, for she went on to tell me: "Now you mustn't do this yet. Watch her crop and if it gets badly distended, I'd be glad to come up and help you." Oh yes, the crop was sewed up with common thread. I asked when such distention would be visible, and she said, "Oh, in a week or so." A week or so! How would I ever get through a week of this worry? She told me to feel the crop about twice a day, and assured me that I would be able to tell.

Her advice did help by giving me something to do. The only person to remain on an even keel was Robert. As for having me feel her crop, she loved it. She always had enjoyed having her breast stroked, and this presented no problem. Neither did the bias binding. The crop never became distended, life went on as usual, and, as usual Robert (and nature) took care of everything.

Several weeks later, a visiting ornithologist told me that I need not have worried at all. He said that a bird's crop can take care of almost anything except stone or metal. I don't recall that he said anything about hydrochloric acid, but there is something in there that eventually liquefies foreign objects. And whatever it is, it did so in this case, to my immense relief. However, I saw to it that such an occasion did not arise again.

She gave me so much to remember. One day I started to make a pie, and had just floured the Formica counter top liberally when the telephone rang. As I talked I was aware of Robert's having found something interesting, as I could hear excited little chirps. She had. She had discovered the flour, and she must have liked the feel of it on her feet—though how she could feel anything through that finely armor-plated foot is beyond me. Such a pattern of quail tracks! By the time I got to her, she had gone on to bigger and better fun: a shallow bowl in which were two cups of sifted flour. In she hopped, and it must have felt like a soft bath to her. Flour flew all over everything, including me, when I grabbed her. I wiped her off with several pieces of Kleenex and started over again with my pie.

Only once did I see her do anything which I thought was done from temper. Not real temper, perhaps, but Robert had certainly been provoked. This time, there was someone with me, who, by the way, did not see any funny side to it at all. It was the mother of a little boy: I had been going over his schoolwork with him, and on this particular day the mother had come by herself to discuss the results. She had brought all his school reports, as I had asked her to, and these were spread out on the table in sequence. Robert was on the table as usual, and two or three times I had gently pushed her away from the papers, explaining to her that those were not for her. She retired to the edge of the table and stood there looking at us. In less time than it takes to write it, I saw her walk up to her glass of V-8, turn her back to us and calculatingly grasp the edge of the glass with her claw and tip the whole thing over the reports.

The woman yelled, "Oh, you bad bird! Look what you've done!" I flew into action, getting things cleaned off and mopped up, and Robert, who had never had a cross word said to her in her life, disappeared. When the interview was over, I escorted the mother to the kitchen door and, peeking in the front bedroom, saw Robert up on the Bible, sulking. She was the picture of a sulk, with feathers hunched up and head down. As soon as she saw the stranger's car pass the window, out she came, as gay as a cricket. All was forgiven. But that time I had to wash the blanket on the Bible, for she left very tangible evidence of having been upset to quite a degree.

While on this subject I want to record that never once, in the three months she was with me, did she ever make a mistake on a shoulder or in anyone's arms. Luck? I don't think so, I think Robert just knew.

Two institutions profited greatly by her visit. One was the Eastman Kodak Company, which must have been able to declare an extra dividend that fall. Robert was such a cooperative subject and the patio such an ideal spot for photography that many people became shutter-happy over her. I watched one woman use up three entire packs of Polaroid film in less than an hour.

The other profit went to the overseas air-mail people. As well as I thought I knew Robert before she came, as well as I thought I was equipped with knowledge of just how to care for her, many things came up demanding a letter to her family, and an immediate answer. For instance, was it all right for her to have cantaloupe seeds? She was avid for them when she first saw them, and I was afraid they were too large for her to handle. Back came an answer saying that it was perfectly safe. I think if Mildred and Tommy had not been such gentle, courteous people they would often have written, "How many times do we have to tell you that Robert knows what is good for her?"

Their first letter to me, mailed on their arrival in London, was just what would be expected. How was she? Was I having any trouble? Was she eating well? They had thought of her almost constantly while on the ship. Another air-mail letter came the next day, after they had received my letter awaiting them at the hotel in London. Such relief. After that, they relaxed and enjoyed the trip. They must have been amused at some of my many minor concerns, but they always answered immediately and reassuringly.

I found myself getting a bit anxious by the time their return was only a week or two away. Everything had gone so well, it was so nearly over, that I asked nothing in the world but to be able to hand Robert over in a state of good health and happiness.

At last the day came, and what a reunion! I am sure she knew them, and she left no doubt whatever that she knew her own home when she was put down on the familiar davenport. At first she just looked around, then she hopped to the floor and made her tour of inspection. All was well.

She took up life and living there exactly as she had left off three months previously—with one exception. She would not take any V-8. Water and orange juice presented no problems, but V-8, which had been almost a staple, and so good for her, she would have nothing to do with. One day when I was over there, and watched her turn away from it when it was offered to her, an idea struck me. I immediately drove back to my house to fetch the little squat glass in which I had always put her

V-8. We set it on the floor and tapped for Robert. She spied that amber glass and ran to it as fast as she could, and drank and drank and drank. From that day it always had to be in that glass.

I had sent the balls of mohair yarn home with her, to replace the red velvet hat she had rejected forever, and she continued to use them as her bed. The garden shoes once more became her place of refuge when she was alone, and all else was as it had been.

She did not visit me again until the following Thanksgiving, just about a year after she had left me. There was not a minute of readjustment, and from then on she had two homes and knew it. She needed a second home occasionally after the grandchildren acquired both a puppy and a cat. She always seemed to enjoy those visits. After all, a change of scene now and then is acceptable to all of us, so why not to Robert? It always makes home seem sweeter on return to it.

11

NBC TELEVISION

I OFTEN RECALL a statement made by a little nine-year-old girl who came with a group of so-called "Fresh Air children" from one of New York City's most underprivileged districts to spend the summer at a home on the beach. The little girl adjusted beautifully to the children in the home, and the other children in the summer colony loved her. She became one of the crowd. It was an ideal summer for her, for she learned to swim, she learned to help on a sailboat, and she was overjoyed to spend a night on a sleep-out on the beach. At the end of her two-month stay, someone asked her what she thought had been the most exciting part of the summer. Without a moment's hesitation, she said, "Oh, the telephone!" "But why the telephone?" "Oh," she said with her eyes glowing, "when it rings, you—well, you just never know!"

My telephone rang one morning in late February, 1964, and I certainly "never knew." The voice was that of a program arranger from NBC in New York. How he happened to get my name,

I'll never know—perhaps from an article in our local paper. After a few questions about Robert (Was she *really* all they had heard?) there came the bombshell question: Could I bring her down to New York to be on television? I was so thunderstruck at the idea that I blurted out the first thing that came to my mind:

"Oh, I would have to talk with her parents. I couldn't give an answer right off."

The voice at the other end burst out laughing, at the word "parents," I guess, and suggested that I talk it over with them and call him back the next morning.

I went over to the Kienzles immediately and dropped the bomb there. Our first reaction was to do it. Of course, it meant that Mildred would have to go with me, as Robert would have to have someone with her in the hotel. We did not worry about her traveling to New York, but if she were left in a hotel room and a chambermaid came in, she would be so delighted to see a companion that—well, anything might happen. But we still thought it might be managed. Then sober second thoughts set in. How would a little quail, accustomed to clear, clean Cape Cod air, react to the soot and general pollution of city air? Would a change of water upset her? Word of our dilemma got around and we had a great deal of advice. Finally, someone experienced in television work gave a most emphatic "no" to it all. She came up with a reason which none of us had thought of. She told us that the heat from the television lights was so intense that it would be extremely dangerous to subject Robert to it. That ended all discussion.

The next day I telephoned NBC and told them the decision. Then they asked if I had or could get some good professional photographs of her and come down myself to be on the "Missing Links" program. I said I was certain I could, and in a few days another call set the date. John Schram came up from the camera shop and spent about an hour with Robert. He worked quietly and so patiently, following her about, snapping promising shots. At last she stood by a low Delft blue bowl in which were a few sprays of silvery-gray petrified twigs. "There," he said, "that's the one," and it was. When we saw the results, about ten beautiful pictures blown up very large, we could hardly choose the four we liked best. The one with the blue bowl was the masterpiece.

I had already been asked to write a few facts about her from which the program director was going to choose five, putting each one on a separate card for me to read aloud on the show but always leaving out the main word. It would be up to the panel to guess what that word was.

Then I carefully packed Robert's one egg in its black-velvet-lined box and off I went to New York.

The whole procedure at the studio was intensely interesting. A dressing room was available for me; chromium carts were rolled around, serving coffee and cakes; all was hubbub and confusion and activity. We had a rehearsal, with a sit-in panel, not the real one, and it was well that we did. I kept forgetting to use the masculine pronoun, as Robert was to be "he" until the dénouement, when I produced the egg. After the rehearsal I had an hour and a half before we really went before the cameras and the audience. The master of ceremonies handed the box containing the egg to an official-looking man with NBC in silver on his shoulders, and told him to have it back on the podium at five minutes past twelve. We were to be on tape, which meant that I would be able to see the show in my own home when it was presented on the air. I was told when that would be, and spent the time between the rehearsal and the recording writing postcards to my friends telling them about it.

At one point a member of the NBC staff who had overheard part of the rehearsal asked me if I really did have a quail's egg there. I assured him that I did, and went to find the man in charge of it so that I could prove it. At least, that is what I thought I was going to do. But that was not the way it worked out. I found the man and made my wants known—and was told firmly:

"I'm sorry, madam, I am responsible for this egg, and I cannot let it out of my hands." And he stuck to it.

Afterward, I wondered if his wife asked him, at the end of the day, how things had gone, and if he told her that he had worked very hard; he had held a quail's egg for an hour and a half.

It all went off well: I remembered my pronouns, the egg was where it was supposed to be, the panel was delightful and guessed well enough so that I won three hundred dollars. So, considering that all expenses were paid by NBC, travel, hotel and all, it was a very pleasant and profitable experience.

Since Robert was well known in town, having once been referred to on a local radio program as "Orleans' first citizen," little groups gathered in many homes the morning we went on the air. It was also on a coast-to-coast hook-up, and soon the letters began to arrive. I had been introduced to the panel and I had mentioned the name Kienzle during the program, but neither name is easy to catch. I had of course mentioned Orleans on Cape Cod; in fact, the master of ceremonies had mentioned it too, and Robert is an easy name. So letters came to *Robert the Quail, Orleans, Mass.* She had appealed to many, many people, even through photographs reproduced on television. Several such letters enclosed dollar bills, asking for copies of the picture postcard of Robert, which had also been mentioned on one of the cards I had read to the panel. Requests for those postcards came from Texas, from Oregon, from Florida. I heard from people I had not heard from or seen for decades. Many of the letters asked for more information about her, and many asked specific questions. Some of the three hundred dollars went for postage, as I answered hundreds of letters from people in twenty-three different states.

We were very thankful that we had not taken Robert to New York to appear in person. While I was in the studio I was told of a very remarkable parrot which had been on the same program six weeks previously and was only then getting over the effects of severe dehydration. A parrot is a big bird compared to Robert, and I shudder to think what might have happened to her five ounces of body weight.

Out of all this correspondence came a revelation. We found that neither Robert nor we were unique in our relationships. Many other quail have been as fortunate as Robert in their adoption of human habitation and mode of life. Some must have private secretaries, since even I cannot believe that a quail can type. But many letters directed to Robert were signed by other quail, and practically constituted a pen-pal club. A correspondence developed between Robert and Dennis, a quail who lives in Bradenton, Florida. We heard from and about Tweety, in Elkton, Maryland; Miss B in Bruceton Mills, West Virginia; clippings were sent telling of a quail in Arizona, whose favorite haunt was a swimming pool, to the delight and astonishment of many visitors. Some letters seemed to have matrimony in mind. One woman thought there were enough of us to warrant the formation of a quail club. The appeal of the *Colinus virginianus* is indeed widespread. (Speaking of names, there is a town on Cape Cod called Waquoit, which, historians tell me, is the early Cape Indians' interpretation of the bobwhite call.)

Several letters were most encouraging because they mentioned ages; one quail had had his seventh birthday. His family regularly employed a baby sitter, as I did while Robert was living with me.

Up to the time of the television publicity, we knew of only one other domesticated quail. This one was also on Cape Cod, and had been rescued from entanglement in a roll of chicken wire during what was probably her first venture into the world. She had been immediately freed and put down near the rest of the brood. However, this was of no help, because the mother would not accept the baby once she had been handled by human beings. This quail, named Bumble, was acquired during the same week that Robert came to us, and was taken care of in much the same way. The development of the two was very similar until they reached the age of a year and a half. Then Bumble developed a real dislike for blonds. Since her family were artists, with many people coming to the house, this posed a problem. They tried putting her in a closed room while strangers, especially blonds, were in the house, but this frustration did nothing to improve the situation. The day came when a decision had to be made. After investigating, they took her to the bird sanctuary and tearfully said good-by to her. They waited several days before calling to see how she was getting along. To their delight, they found that she had adjusted well to the new surroundings, often going to the house to be taken in at night. Visitors walking along the paths in the sanctuary are often astonished to find a little quail accompanying them, sometimes taking advantage of a friendly shoulder. She joined a group of quail, but was never completely accepted. When her adoptive parents went to see her several weeks after leaving her there, Mr. Bailey, the director of the sanctuary, knew where the covey usually fed, so they went there and called her. Bumble came running up to them, got on her foster father's shoulder and talked to him. After a very few minutes she gave a loud chirp and flew off to join the others. Once she got in a car with some young people, but was rescued just before they drove away. The young people had known nothing about Bumble, and I am sure they had not intended to kidnap her.

One summer day she disappeared. A few days later a lady about three miles from the sanctuary was sitting out in her garden and saw one lone quail come out of the shrubbery. She later reported that she believed she could have picked up the bird as it snuggled against her feet. It must have been Bumble.

We are all familiar with delightful stories of wild animals, who, for various reasons, are turned back to the wild after having lived with human beings in homes. We know that these animals, sometimes as long as a year or two after being given their freedom, show unmistakable evidence of recognizing their human friend, or something reminiscent of their previous home. Evidently this behavior is true of birds. As for Bumble, while at the sanctuary she often sought refuge in the Baileys' house on a stormy night, she always asked to go out again in the morning. In contrast to the wild animals, I do not think she was ever completely able to care for herself in the great outdoors. She made an interesting transition. Perhaps she was able—just not willing. Who knows?

12

HE LOVES ME

ROBERT'S VOCALIZING, her many sounds which we could interpret, and the many we could not, were by now taken for granted. But in May, before her second birthday, Mildred and Tommy were conscious that something new had been added. Stated more precisely, it was in the process of being added. Robert would spend several minutes at a time, seeming to practice. They watched her head bob back and forth, and her throat flutter and swell, as she stood motionless, trying to say something. Over and over, this happened, and before long they were sure the end result was going to be "Bob White." Sure enough, it finally came out loud and clear. She appeared to be pleased with herself, and trotted off, repeating "Bob White" several times during that day and occasionally for about a week. Then she stopped and it was never heard again. I am sorry I didn't hear it myself, but several others did, and all attest that it was a fine "Bob White." We heard it often from the outdoor quail, and Robert must have heard it too, but it never awakened any response from her. Only later did we learn that there is no reason to have expected it to stimulate her, as it is not the mating call of the quail.

The following week we had plenty of opportunity to hear the real mating call. A handsome male appeared, with the apparent object—matrimony. Through the window we watched. When Robert was out in the large screened section of the patio busily looking for little bugs, the male would pace up and down the length of the screened side, calling, bringing offerings, his chest expanded and his crest erect. Often only the thickness of the screen separated them. Robert just was not interested. The persistence and daring of that male were unbelievable. He knew what part of the house she was in when she was indoors, and there he would be too, outside the nearest window, singing, calling and showing off his charms. He even flew up to the branch of a tree and called. I am told that this is most unusual for a ground bird like the quail, but he was desperate. Even when Robert was out in the yard with some member of the family, the male kept up the courtship, sometimes coming to within six feet of Mildred or Tommy.

We had often discussed this possibility. In fact we were all agreed that if she did feel the stir of romance and decide to join forces with a mate, we would accept it as the best conclusion and just call it a happy ending. Again, we had figured without Robert. She simply gave her suitor no encouragement, and after two or three weeks of frustration he gave it up as a bad job.

Nature took its course, however, and the following week, on June 9—which happened to be the Kienzles' wedding anniversary—she laid an egg at Mildred's feet. By June 25, she had laid three, two of them being deposited in Tommy's lap. I do not know whether or not this five- or six-day interval is common to wild birds. But that was the way Robert functioned. Someone suggested that a nest be made of soft bits of cloth and that the eggs be placed there to see if the nesting instinct would operate. It didn't. No interest. In a day or two after the third egg, Robert was ill. She threw the family into a panic with her cries of pain. Contrary to habit, holding her did not soothe her. She would settle down for a moment, then, with the distressing cry of pain, fly off in a frenzy. In one of these wild flights she struck her head against a beam in the ceiling, which did not harm her but did not help matters any. She would not eat; she looked miserable and dejected. Nothing distracted her. The second day of this distress, Mildred saw an odd wet spot on the floor. On examination, it proved to be the white of an egg. For the time being, Robert was less uncomfortable, but the Kienzles were not, for they realized that an egg had broken inside her. The shell, before it hits the air, is softer than it is later; but even so, it was inside her. The next day, a normal egg pushed out the broken part and all was well. There was no more trouble after that, and by mid-July she had laid thirteen eggs—and that was the end. Tommy's sister Charlotte was a guest during this egg-laying time, and loved the bird dearly. The last egg was laid at Charlotte's feet as she was packing her suitcase in the bedroom. The first had been an anniversary gift, and the last a farewell gift to a departing guest. While I know that these things were not premeditated or planned, they were nonetheless endearing.

Sometimes she laid an egg on the floor which was not picked up immediately. These she pushed under a bed or davenport to be found later. I often thought of the ornithologist who had told me, when she was only a few months old, that she might leave us during the second mating season of her life. That time had now come and gone and Robert had repudiated her own kind in favor of her human environment.

Any family, be it composed of people only, or of people and pets, must expect minor crises from time to time. Without them, life would be exceedingly dull, since the word "crisis" implies the unexpected. Robert's next minor crisis was certainly most unexpected. Her curiosity and interest in any grocery bag or other package remained insatiable. On this particular afternoon, Mildred came in with an unusual number of packages. None of them contained groceries, and she went immediately to the guest room to put them on the bed. In her haste to get there before the bundles spilled out of her arms, she pushed rather hard against a sliding door, which stuck and then, in response to a very violent shove, gave way suddenly, precipitating her through the doorway. She was conscious of hitting her left hand and, almost in her subconscious, knew that something had dropped. Robert was at her heels, having greeted her as she came in the house, and Mildred saw her swoop down on something with her cry of delight. Mildred managed to get the bundles on the bed, and as she slid her hand out from beneath them, something caught on the coverlet. You have probably guessed what it was; there was an empty setting where a diamond had been. In her engagement ring there was a good-sized stone in the center, with a smaller one on each side. One of the smaller ones was gone. She was sure that the ring had been intact when she picked up the bundles. She called Tommy, asking him to bring a flashlight. As she told him what had occurred, she suddenly knew the answer. Robert! Since the floors were cork, it was easy to look the area over carefully, but they both knew the search was futile. The diamond was in Robert.

Toothpicks were brought into play again, and her droppings were examined minutely. The diamond was never recovered. We know that birds eat gravel to aid in digestion in the crop, and it was nice to think that Robert had the best grinding material in her digestive system that a bird could have. I recently read an article which stated that the sea gull has the most wonderful powers of digestion of any creature. I find this easy to believe, for I have so often watched them on our shores. The best place to observe sea gulls and their eating habits is at our town dump. I enjoy taking a turkey carcass to our dump early in the morning. The dump, by the way, is very well maintained and there is nothing unpleasant about a trip to it. The technique with a turkey carcass is to leave the car door open, take the carcass a few yards away, throw it and run. The gulls come in clouds, and almost before I can get back into the car there is nothing left but bare bones which look as though they had lain in the desert sun for months. I have seen gulls swallow most of a waxed-paper bread wrapper, and they have also been known to swallow fish to which a hook is still attached.

Every spring I spend hours watching the herring crowding up the Brewster herring run to spawn in the fresh-water pond beyond. The gulls have an orgy with the thousands and thousands of herring available. There is always an especially greedy one who makes an entertaining spectacle of himself. The herring is a trout-sized fish and a gull can easily swallow one whole. And there is always one gull who, after swallowing one herring, immediately swallows another. This is quite a load, but time and time again I have watched a gull swallow a third one. By then the gull cannot become airborne, and staggers under the extra weight. When I have gone toward him as though I were going to touch him, he has always managed to regurgitate the third herring and, with some effort, fly away.

But Robert was not a three- or four-pound gull, and while I knew from experience that she could handle two inches of bias binding in her insides, I felt equally sure that she could not digest a diamond. I also had enough faith in her innate ability to take care of herself to be confident that she would manage this time. One or two people made horrible suggestions—suggestions best left to the imagination—as to the procedure to use to retrieve the stone. I reported these to Mildred and Tommy, who reacted violently to them. Meanwhile, the diamond was replaced in the ring with a new stone, and Robert kept hers.

13

ROBERT ASSERTS HERSELF

AS A PSYCHOLOGIST I have always been interested in the ways different individuals react to similar stimuli; for example, hurt feelings. Some sulk; some are angry and dream up imaginary retaliation; some hold the grudge for a long time, while others throw it off immediately; and some are just deeply hurt without showing it.

This is true not only of people but of animals. Poodles, by and large, seem to be delighted when someone comes to call, and always expect recognition. Johnnie Beals, a poodle friend of mine, does not give up easily when ignored. He will get a toy, bring it to the feet of the caller, push it a few times, finally resorting to short barks. He is not so much hurt as mystified by not being noticed. His father, Jackie, the poodle who became so friendly with Robert, reacts very differently. He greets the caller, tail up and wagging, may give a gentle nudge with his nose, and then if there is no response, walks away, tail down, and curls up resignedly on his chair. There he will stay very quietly until the door has closed behind the caller, when once again, he becomes his bouncy self. These friendly dogs justifiably expect friendship in return.

It is not illogical to suppose that similar behavior might be expected from a bird, especially from such an extrovert as Robert. With the one exception of the time when Robert deliberately upset the V-8 juice on the report cards, she had not come up against the problem of being snubbed. To be sure, she reacted somewhat differently to different personalities, and she had definite preferences. If Gladys Taber was present in a group, for example, Robert always singled her out, climbed up on her shoulder and snuggled happily close to her sunshine-colored hair. Robert was a living exponent of the theory that behavior reflects behavior, or that the example set by parents is far more effective than thousands of words of admonition.

What an example of kindness and affection Robert enjoyed! This came not only from the hundreds who came to see her but, more importantly, from the day-after-day, month-after-month and year-after-year atmosphere around her. Even when there were guests in the Kienzle home for any length of time, the pattern was not interrupted. Letters received from Tommy's sister, Charlotte, showed as much real interest in news of Robert as in news about members of the family.

If it had been possible to spoil Robert, it would have been accomplished during the two months' visit of Don and his German bride. The impression Monika had made on Tommy and Mildred during the German visit proved well founded indeed. The tall, gentle blonde girl was a perfect mate for Don. Both share a lifelong sensitive love of animals and all wildlife, so it is small wonder that their love for Robert was immediately reciprocated. I can picture Monika now, knitting in the big chair, with needles flying and Robert happily ensconced on top of her head. (They both planned on a pet as well as children when they were settled, and it was no surprise when their letters from Australia were full of the doings of Murzel, the kitten. Now Murzel shares their love with Trevor, a baby boy.)

Now the Kienzles were looking forward with intense pleasure to a visit from their elder son, Tom. He and his family had often been there, of course, but on those visits the two adorable grandchildren precluded concentration on any one person. How the grandparents enjoyed those times; the children were changing rapidly, becoming charming companions as they grew. The forthcoming visit was to be different. Tom was completing his Ph.D. thesis and needed a few days of quiet and uninterrupted concentration, conditions never found in a household including two lively small children, a dog and a cat. This was to be the first time that the Kienzles had Tom all to themselves since the advent of his younger brother when he was three years old. The study in the guest wing was ready, menus built around Tom's favorite food were planned, and he came.

The two sons are far from being duplicates. Tom, while kindness itself, felt much less emotional involvement than Don did in pets. His children have pets, and they are beautifully cared for and loved. Both he and Nancy admire Robert and appreciate her uniqueness and are good to her. Since Robert always stayed safely under the big bed while the children were there, she was not so much in evidence as at other times. While Tom was there by himself, life for Robert was more normal. She soon learned, after having been gently removed from the table covered with graphs and charts, that the study was off limits for her. She seemed to accept her new status as just another member of the family, and not the most important one; but she was to prove beyond shadow of doubt that this was not entirely to her

liking and that she missed the accustomed and constant attention. But her opportunity came.

The last dinner before Tom's departure was ready. How the Kienzles had enjoyed mealtimes. Tom, temporarily released from his labors, was relaxed and ready for conversation. And such conversation! It was on a completely adult level, and rather high-keyed discussions often developed. Robert was out of it, and she realized it. She had always accepted guests for dinner as the more the merrier. At a very lovely dinner party, for example, given in honor of my birthday, all the guests vied for her attention. But things had changed this time. Mildred and Tommy were subconsciously aware that she was very quiet, staying at her end of the table where her tiny dish of seeds was placed. Occasionally she would advance to the center of things but, receiving no encouragement, would retreat to her own place.

Her manners at the table were exemplary, and since she was accustomed to being offered tidbits, she had never presumed to take anything from a general serving dish. But nothing was forthcoming this time. She walked to the center of the table, looked things over almost as though making a decision, then with wings slightly spread, hopped upon a dish of quivering mint jelly and padded around on it. All that was lacking was an accompaniment of the song "It's nice to beat your feet on the Mississippi mud." Tom, thinking of the jelly, said very, very firmly, "Oh Robert!" Mildred, thinking of those sticky feet, picked her up and retreated to the kitchen with her. There her feet were washed and dried, with Mildred talking to her as a mother might to a child who "didn't mean to do it." The jelly was replaced, Robert was left on the kitchen counter with a bit of lettuce, and the meal resumed. In a few minutes she was back at her own place on the table, more complacent than chastened. Her next attempt to attract attention was more dramatic and also more disastrous. Inch by inch she approached the arena, and before anyone could stop her, she was into the dish of chopped broccoli and butter sauce, where in less time than it takes to tell it she went into her dirt-bath routine, throwing broccoli and butter sauce all over everything and everybody. This time she was not picked up, she was grabbed before she could shake herself, and the clean-up began. She had done a very thorough job, liberally sprinkling table, people, place mats and even the buffet with gobs of broccoli and butter sauce. And as for Robert—she was a mess. Her breast feathers were soaked; the contents of the dish were on her back, her head and even under her wings. She came as close to having a water bath as she ever had, and she spent literally hours afterward cleaning each feather. She worked at it until she was put to bed—in disgrace, for the first time in her life. I am sure the episode did nothing to endear her to Tom.

The whole story was written to Don and Monika, and their comment was characteristic: "What a shame that that entertainment was wasted on Tom. How we wish we had been there."

Coincidence? Maybe. But the fact remains that she had never done such a thing before and that she never did such a thing afterward. Tom left, all was forgiven and, as far as Robert was concerned, forgotten.

How we have changed in these United States in our attitude toward pets. The millions of dollars spent each year on dog and cat food attest to this emphatically. While I have never seen statistics showing figures broken down according to geographical sections, I am willing to wager that the bulk of such spending is in the East and the far West. The Midwest seems slower to change. Perhaps this is because that area is largely agricultural. I spent many of my childhood years there. Lots of my friends had dogs, and they were just dogs. Table scraps sustained them, and at that time any Midwestern veterinarian would have been insulted had an ailing cat been brought to him. His services were exclusively for cattle and horses. Now we see many veterinarians educated to deal exclusively with small animals. In my childhood, and I am sure it is true today, every farm had a farm dog and a barn full of cats. But they were outdoors dogs and strictly barn cats. Beyond having a large pan of milk poured out for them after milking, the cats fended for themselves. As for having a pet sleep at the foot of a child's bed—this was foreign to all local thinking.

Two charming women who were childhood friends of mine in Iowa bought a home here on Cape Cod not too long ago. I am sure that they were perplexed, to put it mildly, at my feeling of responsibility for Robert during her stay with me. They accepted it graciously, but with little understanding. As they made friends here, and they made many, they began to realize that many of these friends were devoted to pets, and that in some cases they were slaves to their pets. These two friends are both intelligent and broadminded people, and told me one day that they had talked it over and decided that they would just have to reorganize their thinking on the subject of pets. They have succeeded pretty well.

14

AN ANXIOUS WEEK

THE FIRST two and a half years of Robert's life had been blessed with good health. To be sure, she had had her share of small accidents, which caused her a day or two of discomfort, but she had not been seriously ill. She had even had minor surgery, which had not bothered her at all, when a small, spur-like growth appeared at the corner of her beak which seemed to be very annoying. She worked at it herself, scratching it hard. After examining it closely, Tommy decided it had better be removed. He felt sure it could be done without any anesthetic, and he was right. Mildred held her in her hands, rubbing the side of her face and head with one thumb. Robert's head drooped over Mildred's fingers, her eyes were closed, and she was certainly relaxed. Tommy laid the edge of the scissors against the spur, and it did not disturb her at all. Then he quickly snipped it off, and she seemed not to have felt it. She was greatly relieved and the scratching stopped.

None of those little upsets had been serious, but on the first of April, 1965, she was sick, very sick. The first day we were aware that she was not herself, she was very quiet, not active at all, and ate very little. It was too early in the season for molting to start, but of course we considered the possibility of the beginning of another ovulation. The next day Mildred watched her as she rather hesitantly approached her seed tray. Much alarmed, Mildred called Tommy:

"Tommy, I don't think she can see!"

They picked her up and, taking her into a good light, realized that this was indeed true. Her eyes did not look right. All that morning she had been shaking her head frequently, often uttering a little cry as she did so. Both eyes were filled with a clear, gelatinous fluid, which formed rapidly and accounted for the constant shaking. In fact, the next time she shook, they could see the tiny droplets fly. She could not even see a good-sized piece of lettuce held before her.

I saw her that night and it was heart-breaking. She knew her way around the house and was on the floor when I entered the room. But there was no rushing to me with cries of greeting. She did come toward me when she heard my voice, but it was all too evident that something was very wrong. I watched her go to the hearth by the fireplace where a glass of water was always kept for her. She walked up to it, going very slowly, and found the glass. On the first attempt to get a drink, her beak hit the near edge of the glass. She tried again with no luck. Finally she walked slowly along beside the glass till it was against her wing. Then, turning her head at an angle, she seemed to see a little from the side of her eye, and in that awkward position managed to quench her thirst. I left for home. I could not endure watching her. My eyes filled too, but with tears.

I am just no good where a sick animal is concerned. I can

pitch in and do anything necessary for a sick person, but I know what I am doing—and, what is more important, the person knows too. But a sick or suffering animal is so helpless, and I am so frustrated by my own helplessness, that my cowardly inclination is to get as far away from it as I can. I am rather ashamed of this weakness, but that is the way it is. The disgraceful truth is that I did not see Robert again for six days. It was all I could do to telephone, I was so fearful of what the news might be.

I knew that Tommy was worried about her, though he was a pillar of strength. At one time he voiced the fear that it might be the delayed result of one of the times she hit the top of her head so hard. He hoped that it had not caused a blood clot, or something of the sort, back of the optic nerve. There was certainly no comfort in that idea.

I telephoned my friend Fran Leach, who is a wizard with sick canaries and other birds, but she could tell me nothing except to say that she would try to see if she could find out anything about eyes in general. She did have a suggestion, and when the Kienzles went to see her she urged strongly that they give her vitamins. They had already thought of that and had emptied a little capsule which contained multi-vitamins, and had tried to give Robert some, but she objected with what strength she had. Mildred tasted the powder and said it was horrible. But Fran had a special vitamin in powder form which she gives birds, tiny monkeys and other small animals in the spring. To a person, this was much more palatable, having a distinctly nutty flavor. The powder adhered to freshly cooked grains of wild rice, and in this way they were able to get it into her.

Feeding her was a major problem. A bird eats a great deal in comparison with its body weight, and they knew they had to keep her strength up. She could not manage her wild-bird seed because she was always selective, and ate only certain of the seeds. Mildred tried filling a tiny glass with the seeds, and getting her beak into it. But this meant that she came up with a mouthful, which is not the way a bird eats. She opened her mouth, trying to get rid of it, and that did not work. She had always had her wild rice from the palm of someone's hand, and had often had it after she was in her bed and after the light was out. I doubt if any half-hour went by during those days without Mildred's getting a few grains into her. This was a lifesaver in the truest sense of the term. It also ensured her getting the vitamins every day in addition to the nourishment.

News of the little bird's illness spread fast. As is usual in cases of illness, friends hesitated to telephone the Kienzles, with the result that my telephone rang very frequently. The innate kindness of people is heartening. Walter Peers, who had built the patio for me when Robert was with me those three months, and who with his wife had moved into my house to take care of her when I had to be in New York for a week, called one evening to inquire about her. I had to say that there was not much that I could tell him in the way of encouragement, and I thanked him for calling. To that, he replied, "Well, we got to know Robert pretty well, and when you know her . . ." and his voice choked up.

And so it went, long day after long day. Tommy was able to obtain some special eyedrops, which were put in two or three times a day. This remains the only thing done to her, or for her, to which she objected. How she hated those eyedrops. Perhaps they smarted, though Tommy did not think so. He did think that sunshine would be good for her, but the weather did not cooperate. We had the rainy and windy days which April so often brings, and outdoor exposure for Robert was out of the question.

About the fourth day the sun did come out. Mildred carried Robert out to a sheltered part of the patio and, shielding her with her coat, let the sun shine on her back for a few minutes. The next day was even better, and Robert was put down in a sunny spot. Surprisingly, she took a dirt bath, a rather feeble one, but a dirt bath. It must have felt good to her, for she lay on her side against a little mound of dirt for several minutes. The bath did not help the top of her head, which needed cleaning badly. The topknot feathers were stiff and crusty from the droplets that had landed there when she shook them from her eyes. Washing it with water was out of the question, but petroleum jelly rubbed in helped temporarily.

We knew some food was going through her, but we also knew that she was getting far from enough to maintain her very long. On the seventh morning she lay quietly, perhaps resignedly, in Mildred's hand, but when the eyedrops were put in, she put up an amazingly strong objection, wriggled out of the hands, down onto the floor, and made a beeline for her feeding tray. Scarcely believing their own eyes, Mildred and Tommy watched her as she selected her favorite seeds, and heard her uttering her little contented chirps. She could see! Then she headed for the living room to a pan of growing chickweed which two of her admirers had found under a sheltered branch in their garden, but which till this time she had not been able to eat. She had tried, swallowing at random whatever she found—tiny twigs, bits of leaves and occasionally a bit of the green. Now she hopped up in the pan, and how she ate!

Mildred called me immediately, and the world brightened. For several more days a slight amount of the fluid came in her eyes, but the drops were diminished to once a day, to Robert's relief. For several days she tired easily, asked to be put to bed very early and spent a lot of time on the kitchen counter under her lamp, sleeping on her side. What had been the trouble? A cold? A virus infection? Bird pneumonia? Who knows. What had cured her? Eyedrops? Sunshine? Vitamins? Prayers? Who knows. What did it matter? She was well once more.

We often hear it said that a bird eats its own weight in food every day. For the few days after the return of her sight, Robert must have done just that.

It often takes an illness to make us appreciate health.

A very considerate veterinarian once told a friend whose dog was sick that she must remember that animals do not worry, that they have no fear as to what may be ahead when they are sick and that they just take each day as it comes. This is doubtless true of birds as well as animals, and it is a comforting thought. Apparently human beings are the only creatures cursed, with the ability to worry, and we certainly worried. By the same token, maybe we are the only ones blessed with the capacity for appreciation, and we—all of us who were so deeply concerned—appreciated Robert's return to health and happiness.

15

THE LAST CHAPTER

ROBERT approached her fourth autumn season in good spirits. Once again the patio was delightful with fallen leaves under which she knew there were good bugs and spiders. It meant a great deal to me to have her well because she was soon to spend another ten days at my home while the Kienzles were going to Lexington to help in welcoming a new baby in their son Tom's family. Since Robert was as much at home in my house as in her own, I did not give the adjustment a thought this time. Perhaps I took too much for granted. I believe Robert knew that I would bow to her every whim.

The very first night, when she signified that she was ready to be put to bed, she insisted on a change of routine. Everything was just as it had always been on other visits, but something was not to her liking. I held her a few minutes, talked to her and then put her up on her shelf on her own ball of yarn. I closed the door and settled down to doing the many things I had not done during the day. Almost immediately I heard her fly down, and knew she was standing at the bedroom door trying to get out. Of course I went in to her, and we tried again. After four futile attempts to get her to settle in for the night, I was at my wits' end, when suddenly I thought of the little Swiss music box. Once more I put her on her shelf, but this time I brought the music box in with me and sat on the bed and wound it up. It plays two minutes at a winding, and during that time there was not a sound from Robert. But as soon as it ran down, little interrogatory chirps came from the shelf. I wound it again. Again silence until it had run down. After the third playing—silence. I wound it once more, muffled it with a blanket at the foot of the bed and tiptoed out. There was not a sound until nine o'clock the next morning. The whole performance was repeated that night, and I gave in. Toward the end of her visit I tried it again without the music box, but had to go back to it. When she returned home, I told the Kienzles about it, and the first night they tried their music box, which is a duplicate of mine. Robert objected very vocally, and did not want it. It really seemed that she had taken advantage of me, but I was more than willing.

She made one other change during this visit. She discovered that by flying up on a little chest which stands under a window of the keeping room she could not only look out to her solarium but ask to go out that way. Never again did she stand at the door as she had always done. Also this time, when I opened the window she would go out by herself and stay for long periods of time. Only when I heard her little lonesome cry did I go out. She would not come in through the window, but she always went out that way.

She had many visitors, as usual, most of whom knew her well. Several people remarked that she seemed to be slowing down. I would not agree to that, but did have to admit that she seemed more mature, more sedate and less mischievous. Perhaps she did spend more time sitting on my shoulder as I wrote, and I had to admit that she no longer interfered with my typing. She was very lively out of doors, and I often saw her jump up to a height level with the middle of the window to catch a bug. I look back on those ten days with great pleasure.

Early in November the little spur-like growth at the corner of her beak reappeared. It had been successfully snipped off several times, but now it began to spread inside her mouth. Whether the outer growth interfered with her seeing her food or whether an actual failing of sight was the cause, we did not know, but she required a great deal of assistance in eating. The growth prevented her from completely closing her beak, and she could not preen each individual feather as she had always done so carefully; but she tried. Soon every seed, every bit of food, had to be put well back in her mouth, and Mildred spent hours each day doing so. She always saw to it that Robert went to bed with a nicely filled crop. I doubt if any pet, or for that matter many persons, ever had such devoted care. Even water had to be given to her.

Through all this her spirits and her disposition did not change. She even found a new delight, one which was quite foreign to the nature of her kind. She took several flights up to the very peak of the roof, and there she would stand, on the ridgepole, stretched to her full height of ten inches, in a tiptoe position, just looking around. Quail will on very infrequent occasions seek refuge on a low limb of a tree, but they are not given to high perching. Robert seemed to like the enlarged view of the world, and sometimes stayed there for fifteen minutes.

During those weeks more and more frequently she would

leave her shelf early in the morning, and Tommy would find her cuddled under his chin when he awoke. She almost always chose Tommy.

One evening, returning from a convention in Boston, Mildred told Tommy about a lecture which had impressed her. The theme had been that, instead of wishing for powers we do not have, we should do all the good we can with what we have at hand. The speaker had illustrated his point by two references—one from Exodus when Moses, discouraged and downhearted, heard God ask, "What is that in thine hand?" Of course it was the rod, which he struck upon the ground and by means of which the children of Israel were freed; the other being Lincoln, equally downhearted, who heard a voice say to him, "What is that in thine hand?" Of course it was a pen, and history has recorded the lasting power of that pen. Mildred continued to think about it as she went to sleep. In the morning she woke with her hand on the pillow, palm upward, to find Robert asleep in it. The thought flashed through her mind, "What is this in mine hand?" Not the rod of Moses, not the pen of Lincoln, but, in her own small way, quite a power for good. More than one young man, after seeing and knowing Robert, had vowed never to shoot a quail again. Many avenues had been opened leading to interest in and even study of birds and birdlife. And hundreds of people had been entertained and amused by her.

At Thanksgiving, friends in Wellfleet had as guests a man and his wife from Philadelphia who had heard and read so much about Robert that they were very anxious to see her. At that time I had not seen her for several days and was doubtful that I ever would again because she had become so weak. But I explained the situation to Mildred and Tommy, who said I might bring them if I prepared them for seeing her not at her best. I agreed, and the day after Thanksgiving we went, after

I did all I could to see that they would not be disappointed. I might as well have saved my breath.

Robert rose to the occasion, greeted them with her unique welcome sounds, chirped conversationally to Mr. Johnson as he held her and, for the last time, was the perfect hostess. In spite of what I had told them, the visitors thought she was beautiful. Her feathers, with their amazing design, looked uncared for to us, but not to the guests. They signed her guest book, one of them commenting after her signature, "A joy to hear and see." So ended her book except for a tribute later written by Tommy.

The morning of December 2, Tommy found her again under his chin, and let her stay there for some time. She seemed so contented, and so tired. However, she accompanied him into the bathroom as usual, and hopped up on the breakfast table where she was fed. Later that day she cocked her head at the sound of some soft music, flew up on the back of the davenport and swayed and sang with the music as she had not done for several weeks. She wanted to be held a great deal that afternoon, and was put on her shelf early in the evening. There, for the first time in her three and a half years of life, she tucked her head under her wing, and was immediately asleep.

The next morning Mildred had to go to a meeting, and she called me asking if I would take her, adding that Tommy didn't want to leave Robert, who was still asleep, still with her head under her wing, but definitely sleeping. After I took Mildred home, I bought a steel strongbox, which I lined with soft wool nylon blanket material and left on their doorstep without going in. After lunch my telephone rang. I knew as soon as I heard Mildred's voice what she had to tell me. She said, "I just want to say one sentence. Robert went finally to sleep with her head still under her wing."

I could make no reply.

She lies in the little box, in the same position in which she slept, in the section of the patio where she had loved to take her sunbaths and hunt for bugs. She is watched over by a pair of hand-carved stone quail from Japan.

I often think of Mildred's remark of three years ago when we wondered so many things, among them what the end would be, and she said, "I know one thing, whatever it is—oh how we'll miss her." And we do.